Biological Networks

Biological Networks

Special Issue Editors

Rudiyanto Gunawan
Neda Bagheri

MDPI • Basel • Beijing • Wuhan • Barcelona • Belgrade

MDPI

Special Issue Editors
Rudiyanto Gunawan
Institute for Chemical and Bioengineering
Switzerland

Neda Bagheri
McCormick School of Engineering Northwestern University
USA

Editorial Office
MDPI
St. Alban-Anlage 66
4052 Basel, Switzerland

This is a reprint of articles from the Special Issue published online in the open access journal *Processes* (ISSN 2227-9717) from 2017 to 2018 (available at: https://www.mdpi.com/journal/processes/special_issues/biological_networks)

For citation purposes, cite each article independently as indicated on the article page online and as indicated below:

LastName, A.A.; LastName, B.B.; LastName, C.C. Article Title. *Journal Name* **Year**, *Article Number*, Page Range.

ISBN 978-3-03897-433-8 (Pbk)
ISBN 978-3-03897-434-5 (PDF)

Contents

About the Special Issue Editors

Rudiyanto Gunawan, PhD, is an Associate Professor in the Department of Chemical and Biological Engineering at University at Buffalo, where he directs the Chemical and Biological Systems Engineering Lab (CABSEL). He received a BS degree in Chemical Engineering and Mathematics from the University of Wisconsin-Madison and MS and PhD degrees in Chemical Engineering from the University of Illinois Urbana-Champaign. Dr. Gunawan is engaged in a broad range of activities in the areas of computational systems biology, bioprocess engineering, and bioinformatics. The mission of his research is to create enabling and innovative tools for the extraction of mechanistic and actionable insights from biological data, based on rigorous mathematical underpinnings, systems modeling and analysis, and advanced machine learning and optimization algorithms. These tools have been applied to a range of problems from the biopharmaceutical, pharmaceutical, and biomedical fields. Dr. Gunawan has published more than 75 journal articles, conference proceedings, and book chapters in the areas of systems biology, bioprocess engineering, bioinformatics, biology, and biogerontology. He has given more than 35 invited and keynote lectures at academic research institutions and regional and international conferences. Dr. Gunawan is a co-recipient of two Best Paper awards from the *Journal of Process Control and Computers* and the *Chemical Engineering* journal

Neda Bagheri, PhD, is an Assistant Professor in the Department of Chemical and Biological Engineering at Northwestern University, where she directs the Modeling Dynamic Life Systems (MoDyLS) Lab. She received BS, MS, and PhD degrees in Electrical and Computer Engineering from the University of California Santa Barbara. Dr. Bagheri's research lies at the cutting-edge intersection of engineering and biology to solve exciting challenges within medicine and basic science. Her group is particularly interested in challenges related to identifying engineering design principles that underlie, explain, and rationalize complex biological function, as well as understanding how extrinsic factors can be used to optimize therapeutic interventions. To accomplish these goals, her interdisciplinary team of engineers, basic scientists, and applied mathematicians combine experimental data with novel computational strategies derived from statistical analysis and control theory to attack problems from creative angles not possible with single-discipline methods. Solutions to these challenges have the potential to address important problems in cancer and immune system diseases and to uncover new fundamental understandings of microbial and circadian biology. In recognition for her research accomplishments, Dr. Bagheri was awarded a CAREER Award from the National Science Foundation (NSF) in 2017. Her research has been published in high-profile journals, including the *Proceedings of the National Academy of Sciences* and *PLOS Computational Biology*. She serves on scientific advisory boards for an NSF Science & Technology Center, Immuneering Corp., and the Allen Institute for Cell Science. She is recognized internationally for her leadership in the fields of computational and systems biology.

MDPI

Editorial

Special Issue on "Biological Networks"

Rudiyanto Gunawan [1,*] **and Neda Bagheri** [2,*]

1   Department of Chemical and Biological Engineering, University at Buffalo, 302 Furnas Hall, Buffalo,
    NY 14260, USA
2   Department of Chemical and Biological Engineering, Center for Synthetic Biology, NSF-Simons Center for
    Quantitative Biology, Northwestern Institute for Complex Systems, Northwestern University,
    Technological Institute, 2145 Sheridan Road, E154, Evanston, IL 60208, USA
*   Correspondence: rgunawan@buffalo.edu (R.G.); n-bagheri@northwestern.edu (N.B.);
    Tel.: +1-716-645-0952 (R.G.); +1-847-491-2716 (N.B.)

Received: 20 November 2018; Accepted: 20 November 2018; Published: 27 November 2018

Networks of coordinated interactions among biological entities govern a myriad of biological functions that span a wide range of both length and time scales—from ecosystems to individual cells, and from years (e.g., the life cycle of periodical cicadas) to milliseconds (e.g., allosteric enzyme regulation). For these networks, the concept of "the whole is greater than the sum of its parts" is often the norm rather than the exception. Meanwhile, continued advances in molecular biology and high-throughput technology have enabled a broad and systematic interrogation of whole-cell networks, allowing for the investigation of biological processes and functions at unprecedented breadth and resolution, even down to the single-cell level. The explosion of biological data, especially molecular-level intracellular data, necessitates new paradigms for unraveling the complexity of biological networks and for understanding how biological functions emerge from such networks. These paradigms introduce new challenges related to the analysis of networks in which quantitative approaches such as machine learning and mathematical modeling play an indispensable role. The Special Issue on "Biological Networks" showcases advances in the development and application of in silico network modeling and analysis of biological systems. The Special Issue is available online at: https://www.mdpi.com/journal/processes/special_issues/Biological_Networks.

Identifiability and Design of Experiments for Biological Network Models

A well-known challenge in the development of computational models of biological networks is the identifiability of model parameters. A model is said to be structurally identifiable when its parameters can, in principle, be extracted from measurements of the output responses of the model. The lack of structural identifiability implies that any attempt to determine model parameters from measurement data is futile. The paper by Villaverde and Banga [1] focuses on assessing structural identifiability by using the concept of observability. More specifically, for certain initial conditions, structurally unidentifiable models can mistakenly be ascertained to be identifiable. The paper provides illustrations of such challenges through biochemical model examples and proposes a procedure to overcome this complication.

Even in cases where a model is structurally identifiable, the accuracy of parameter estimates can remain poor. A concept related to structural identifiability is *practical identifiability*, which addresses concerns of parameter uncertainty. A key factor that controls parameter uncertainty is the design of experiments. Two papers in this Special Issue explore the model-based optimal design of experiments (MBDOE) via the Fisher information matrix (FIM). The paper by Sinkoe and Hahn [2] showcases the importance of optimizing experiments for improving the practical identifiability of model parameters, especially in connection to dynamic biological data and modeling. More specifically, the paper describes the application of the FIM-based D-optimality criterion and the Morris method for computing parametric sensitivities, to optimize dynamic input functions to the interleukin-6 signaling model.

In silico implementations of the optimal input functions show great promise in significantly reducing parametric uncertainty.

A high degree of model nonlinearity (curvatures) can negatively affect the performance of FIM-based experimental designs. The paper by Manesso et al. [3] addresses this issue by introducing a new multi-objective optimization (MOO) framework. This framework identifies Pareto optimal experiments that balance maximizing the information content of experimental data through the FIM with minimizing model curvatures. A proof of concept using a biochemical network model of baker's yeast fermentation illustrates the benefits of using the proposed MOO MBDOE over a number of FIM-based optimal designs and other experimental designs that also consider model curvatures.

Dynamic Biological Network Modeling

Two papers in the Special Issue present new biological network models that span multiple scales. The paper by Ruggiero et al. [4] presents a dynamic model of tuberculosis (TB) granuloma activation describing host-TB pathogen interactions. The model captures the local immune system response to *Mycobacterium tuberculosis*, the dynamics of matrix metalloproteinase-1 and collagen in granuloma, and the leakage of bacteria from granuloma. By changing the values of parameters in the model, the authors are able to assess how perturbations in the immune response (as well as HIV co-infection) affect granuloma activation.

The paper by Lee et al. [5] looks at the modeling of NF-κb signaling dynamics induced by lipopolysaccharide (LPS) in the presence of a cytokine secretion blocker. Parameter estimation based on average single-cell flow cytometry data points to a previously unidentified action of the cytokine secretion blocker, which was validated in additional experiments and subsequent model refinement. The iterations between computational modeling and experimental design highlight how the process of inferring biological networks can lead to new testable hypothesis and insights.

Network-Based Biological Systems Analysis and Optimization

Biological network models enable a systematic and comprehensive analysis and optimization of biological systems, as showcased by three papers in the Special Issue. The paper by Perumal and Gunawan [6] introduces a new dynamic sensitivity analysis that can account for cellular heterogeneity. The analysis, termed molecular density function perturbation (MDFP), introduces time-dependent in silico perturbations to the molecular concentrations of biological species in the network. The application of MDFP to a mathematical model of programmed cell death signaling stimulated by a tumor necrosis factor ligand points to key events in the signaling pathway that determine the cell-to-cell variability in the response to the stimulus.

The paper by Widiastuti et al. [7] outlines a model-driven strategy to estimate and improve the production capability of microbes based on an in silico analysis of genome-scale metabolic networks. The strategy explores the use of *Zymomonas mobilis* to produce succinic acid. The genome-scale metabolic network model enables a combinatorial deletion analysis, leading to the identification of four gene deletions that would amplify succinic acid molar yield by 15 times.

The review paper by Faraji and Voit [8] focuses on the metabolic modeling of crop science with a specific focus on bioenergy crops. In comparison to microbes and animal cells, mathematical modeling of plant metabolisms is still in its infancy, but is expected to become a standard tool in the future. The paper delves into unique challenges and constraints in modeling plant metabolic networks, as well as limitations and mitigating strategies in using popular modeling formalisms to capture the physiological characteristics of plant systems. A case study involving lignin biosynthesis in switchgrass illustrates how mathematical modeling can serve as a powerful tool for strain improvement through the generation of a library of virtual strains.

Network-Based Biological Data Analytics

Biological networks are crucial in the interpretation and analysis of biological data. The paper by Vargason et al. [9] demonstrates how univariate (one variable at a time) statistical analysis is often suboptimal as it does not account for the correlation of data structure arising from an underlying biological network. By using clinical data for autism spectrum disorder as case studies, multivariate analyses were demonstrated to be much more efficacious than univariate approaches.

The paper by Padmanabhan et al. [10] highlights how the network of cellular pathway crosstalk can provide better biomarkers with improved diagnosis and prognosis accuracy. The work presents a procedure to construct a cellular pathway crosstalk reference map, by combining information on chemical, genetic and domain interactions and transcription factors. The reference map is personalized by utilizing each patient's single nucleotide polymorphisms. In the application to the Alzheimer's disease (AD) dataset from the Alzheimer's Disease Neuroimaging Initiative, the authors show how using the patient-specific cellular pathway crosstalk as an additional feature significantly improves the accuracy in assessing the risk of mild cognitive impairment progression to AD.

Funding: There are no funding supports.

Conflicts of Interest: The authors declare no conflicts of interest.

References

1. Villaverde, A.; Banga, J. Structural Properties of Dynamic Systems Biology Models: Identifiability, Reachability, and Initial Conditions. *Processes* **2017**, *5*, 29. [CrossRef]
2. Sinkoe, A.; Hahn, J. Optimal Experimental Design for Parameter Estimation of an IL-6 Signaling Model. *Processes* **2017**, *5*, 49. [CrossRef]
3. Manesso, E.; Sridharan, S.; Gunawan, R. Multi-Objective Optimization of Experiments Using Curvature and Fisher Information Matrix. *Processes* **2017**, *5*, 63. [CrossRef]
4. Ruggiero, S.; Pilvankar, M.; Ford Versypt, A. Mathematical Modeling of Tuberculosis Granuloma Activation. *Processes* **2017**, *5*, 79. [CrossRef]
5. Lee, D.; Ding, Y.; Jayaraman, A.; Kwon, J. Mathematical Modeling and Parameter Estimation of Intracellular Signaling Pathway: Application to LPS-induced NFκB Activation and TNFα Production in Macrophages. *Processes* **2018**, *6*, 21. [CrossRef]
6. Perumal, T.; Gunawan, R. Elucidating Cellular Population Dynamics by Molecular Density Function Perturbations. *Processes* **2018**, *6*, 9. [CrossRef]
7. Widiastuti, H.; Lee, N.; Karimi, I.; Lee, D. Genome-Scale In Silico Analysis for Enhanced Production of Succinic Acid in Zymomonas mobilis. *Processes* **2018**, *6*, 30. [CrossRef]
8. Faraji, M.; Voit, E. Improving Bioenergy Crops through Dynamic Metabolic Modeling. *Processes* **2017**, *5*, 61. [CrossRef]
9. Vargason, T.; Howsmon, D.; McGuinness, D.; Hahn, J. On the Use of Multivariate Methods for Analysis of Data from Biological Networks. *Processes* **2017**, *5*, 36. [CrossRef] [PubMed]
10. Padmanabhan, K.; Nudelman, K.; Harenberg, S.; Bello, G.; Sohn, D.; Shpanskaya, K.; Tiwari Dikshit, P.; Yerramsetty, P.; Tanzi, R.; Saykin, A.; et al. Alzheimer's Disease Neuroimaging Initiative Characterizing Gene and Protein Crosstalks in Subjects at Risk of Developing Alzheimer's Disease: A New Computational Approach. *Processes* **2017**, *5*, 47. [CrossRef]

processes

MDPI

Article

Structural Properties of Dynamic Systems Biology Models: Identifiability, Reachability, and Initial Conditions

Alejandro F. Villaverde * and Julio R. Banga

Bioprocess Engineering Group, IIM-CSIC, Vigo 36208, Spain; julio@iim.csic.es
* Correspondence: afvillaverde@iim.csic.es; Tel.: +34 986231930 (ext. 860234)

Academic Editors: Rudiyanto Gunawan and Neda Bagheri
Received: 12 May 2017; Accepted: 31 May 2017; Published: 2 June 2017

Abstract: Dynamic modelling is a powerful tool for studying biological networks. Reachability (controllability), observability, and structural identifiability are classical system-theoretic properties of dynamical models. A model is structurally identifiable if the values of its parameters can in principle be determined from observations of its outputs. If model parameters are considered as constant state variables, structural identifiability can be studied as a generalization of observability. Thus, it is possible to assess the identifiability of a nonlinear model by checking the rank of its augmented observability matrix. When such rank test is performed symbolically, the result is of general validity for almost all numerical values of the variables. However, for special cases, such as specific values of the initial conditions, the result of such test can be misleading—that is, a structurally unidentifiable model may be classified as identifiable. An augmented observability rank test that specializes the symbolic states to particular numerical values can give hints of the existence of this problem. Sometimes it is possible to find such problematic values analytically, or via optimization. This manuscript proposes procedures for performing these tasks and discusses the relation between loss of identifiability and loss of reachability, using several case studies of biochemical networks.

Keywords: identifiability; controllability; reachability; observability; parameter estimation; nonlinear systems; differential geometry

1. Introduction

The study of parametric identifiability is a fundamental task of system identification, which can be approached from structural or practical viewpoints [1–3]. Practical identifiability analysis aims at characterizing the uncertainty in parameter estimates taking into account the deficiencies in the data used for model calibration. Structural identifiability is a prerequisite for practical identifiability, and seeks to establish whether the model parameters can be uniquely determined from observations of the input-output behaviour of the model—that is, from the model equations only. The concept of structural identifiability was coined in [4] and initially introduced in the context of linear systems. It was soon extended to the nonlinear case and many methods were subsequently developed for its study [3,5,6]. In parallel, the classic dual concepts of observability and controllability were also extended from linear [7] to nonlinear [8]. The relationships between these properties make it natural to study structural identifiability with the tools of nonlinear observability [9]. Indeed, it is possible to check simultaneously the observability and structural identifiability of a model by including its parameters in the state variables vector, and calculating the rank of the resulting (augmented) observability matrix. Several computational methods that build on this idea have been presented in the last decade [10–13].

However, the results of these methods should be taken with caution. As emphasized by Denis-Vidal et al. [14], identifiability may depend on the initial conditions. Saccomani et al. [15] noted that differential algebra methods can incorrectly classify an unidentifiable model as identifiable for certain initial conditions (a detailed distinction between "structural", "geometric", and "algebraic" identifiability can be found in [16]).

Here we show that the differential geometry approach (which assesses local structural identifiability by checking the rank of the augmented observability matrix) has in principle the same limitation as the differential algebra approach with regard to initial conditions. When the identifiability matrix is computed with the numerical values of certain problematic initial conditions, its rank decreases with respect to the general case. However, this situation cannot be detected with symbolic rank calculations. When the system evolves from such initial conditions it may become impossible—depending on the particular case—to determine some of its parameters, which would nevertheless be structurally identifiable if the system was started at a different state. In [15] loss of reachability was identified as the cause of this loss of identifiability. However, other alternative causes may also exist. To assess this possibility we propose to check if the identifiability rank condition is satisfied in two ways: for the generic case (symbolically), and in the vicinity of the initial conditions of interest (by specializing its states to those specific values). By performing the rank test both for generic and particular state values we obtain a more complete diagnosis. Then, if a certain numerical state vector leads to an incomplete identifiability rank, it may indicate that some state variables are no longer reachable from it, which in turn may make the associated parameters unidentifiable. This situation can be checked by assessing the reachability of the system both for generic and particular states—or, alternatively, by simulating the model using the state vector as initial condition. This procedure allows to determine precisely if the parameters are structurally locally identifiable for those values. An alternative solution could be to check identifiability in a range of parameter and state values, as proposed by [11], although the computational complexity of such test makes it infeasible for medium or large-scale systems. Given that, in general, the identifiability condition may be misleading when evaluated for generic values of the states, an important question is whether it is possible to determine the specific values that violate this test. For small models it may be possible to find them analytically by inspection of the system equations. In more complex cases an alternative is to search for such values via optimization.

The organization of this paper is as follows: in Section 2 we provide a compact presentation of the necessary background on structural identifiability, observability, and reachability (controllability) of nonlinear systems, and of the relations between these concepts. When available, necessary and/or sufficient conditions under which these properties hold are given. This is presented using the differential geometry formulation. Then in Section 3 we illustrate with six example models several situations and issues that can appear in relation to loss of structural identifiability, and how this relates to observability and reachability. We begin in Section 3.1 by remembering the known fact that reachability is neither necessary nor sufficient for identifiability. Then in Section 3.2 we show that identifiability rank tests can sometimes (i.e., for certain state vectors) be misleading. We further remark, in Section 3.3, that the knowledge of initial conditions of unmeasured states can improve the identifiability. In Section 3.4 we propose a procedure that assesses whether the identifiability rank test is being misleading, and in Section 3.5 we propose an optimization-based procedure to find particular state vectors will cause loss of identifiability. Finally, we summarize the Conclusions in Section 4.

2. Background: Nonlinear Observability, Reachability, and Identifiability

2.1. Notation and Differential Geometry Concepts

We consider a general class of nonlinear time-invariant systems modelled as a structure M with the following dynamic equations:

$$
M : \begin{cases} \dot{x}(t) = f(x, p, u) = f_1(x, p) + f_2(x, p) \cdot u(t) \\ y(t) = h(x, p) \\ x_0 = x(t = 0, p) \end{cases}
\tag{1}
$$

where f, f_1, f_2, and h are vector functions, $p \in \mathbb{R}^q$ is a real-valued vector of parameters, $u \in \mathbb{R}^r$ is the input vector, $x \in \mathbb{R}^n$ the state variable vector, and $y \in \mathbb{R}^m$ the output or observables vector. The dependency of f, f_1, f_2, and h on the parameters p will be usually dropped for ease of notation. The following paragraphs define several differential geometry concepts that will be used throughout this paper.

Given a smooth function $z(x)$ and a vector field $v(x)$, the *Lie derivative* of z with respect to v is:

$$
L_v z(x) = \frac{\partial z(x)}{\partial x} v(x)
\tag{2}
$$

where $\frac{\partial}{\partial x} z(x)$ is a row vector containing the partial derivatives of the smooth function $z(x)$. In the present work, $z(x)$ can be either the m-dimensional vector function $h(x)$ (when studying observability) or the r-dimensional vector function $f_2(x)$ (when studying controllability). For a k-dimensional function z and an n-dimensional vector x and function v, $\frac{\partial}{\partial x} z(x)$ is a $k \times n$ matrix, and $L_v z(x) = \frac{\partial z(x)}{\partial x} v(x)$ is a $k \times 1$ column vector. Higher order Lie derivatives can be defined recursively as:

$$
\begin{aligned}
L_v^2 z(x) &= \frac{\partial L_v z(x)}{\partial x} v(x) \\
&\cdots \\
L_v^i z(x) &= \frac{\partial L_v^{i-1} z(x)}{\partial x} v(x)
\end{aligned}
\tag{3}
$$

Given two vector fields $v_1(x)$, $v_2(x)$, their *Lie bracket* is the vector field defined by

$$
[v_1, v_2] = \frac{\partial v_2}{\partial x} v_1 - \frac{\partial v_1}{\partial x} v_2
\tag{4}
$$

A $k-$dimensional *distribution* Δ on X is a map which assigns, to each $x \in X$, a $k-$dimensional subspace of $\mathbb{R}^n$ such that for each $x_0 \in X$ there exist an open set $U \subseteq X$ containing x_0 and k vector fields $f_1, \ldots, f_k$, such that

1. $\{f_1(x), \ldots, f_k(x)\}$ is a linearly independent set for each $x \in U$.
2. $\Delta(x) = \mathbf{span}\{f_1(x), \ldots, f_k(x)\}, \forall x \in U$ [17].

2.2. Observability

Conceptually, a system is observable if for each state there exists at least one input which allows to discriminate between this state and all nearby states, by measuring the output [17]. More formally, two states $x_0 \neq x_1$ are said to be *distinguishable* when there exists some input $u(t)$ such that $y(t, x_0, u(t)) \neq y(t, x_1, u(t))$, where $y(t, x_i, u(t))$ denotes the output function of the system for the input $u(t)$ and initial state $x_i (i = 0, 1)$. The system is said to be *(locally) observable* at x_0 if there exists a neighbourhood N of x_0 such that every other $x_1 \in N$ is distinguishable from x_0.

2.2.1. Linear Observability

Before proceeding to nonlinear systems, let us recall that a linear system of the form

$$\begin{aligned} \dot{x} &= A \cdot x + B \cdot u \\ y &= C \cdot x \end{aligned} \tag{5}$$

is observable if and only if rank$(\mathcal{O}) = n$, where $\mathcal{O}$ is the linear observability matrix [7],

$$\mathcal{O}(x) = \begin{pmatrix} C \\ C \cdot A \\ C \cdot A^2 \\ \vdots \\ C \cdot A^{n-1} \end{pmatrix} \tag{6}$$

2.2.2. Nonlinear Observability

The extension to the nonlinear case is straightforward. For a system given by (1), the nonlinear observability matrix is built using Lie derivatives as follows:

$$\mathcal{O}(x) = \begin{pmatrix} \frac{\partial}{\partial x} h(x) \\ \frac{\partial}{\partial x} (L_f h(x)) \\ \frac{\partial}{\partial x} (L_f^2 h(x)) \\ \vdots \\ \frac{\partial}{\partial x} (L_f^{n-1} h(x)) \end{pmatrix} \tag{7}$$

Theorem 1. Observability Rank Condition (ORC). *If the system M given by (1) satisfies rank$(\mathcal{O}(x_0)) = n$, where $\mathcal{O}$ is defined by (7), then M is locally observable around x_0 [8].*

The ORC is a sufficient and "almost necessary" condition for observability [17], meaning that if M is locally observable around x_0, then rank$(\mathcal{O}(x_0)) = n$ for all the states belonging to an open dense subset of the state space. Note that, in passing from linear to nonlinear, the rank condition has changed from being sufficient and necessary to being "sufficient and almost necessary".

Two terms related to observability are detectability and reconstructability. Detectability is a similar notion to observability, but slightly weaker: a system is detectable if all the unstable modes are observable. Reconstructability is also similar, but it refers to the ability of determining the present state of a system from past and current (as opposed to future) measurements.

2.3. Controllability and Reachability

While observability studies whether it is possible to reconstruct the internal state x of a model by observing its output y, controllability asks whether it is possible to control x by manipulating its input u. Controllability and reachability are subtly different concepts: in *reachability* the question is which states $x(t_f)$ can be reached in finite time from the initial state, $x(t_0)$, which is fixed. In *controllability* the question is which states $x(t_0)$ can be driven to a final state, $x(t_f)$, which is fixed. A more precise definition of reachability is as follows: a system (1) is said to be (locally) *reachable* around a state x_0 if there exists a neighbourhood N of x_0 such that, for each $x_f \in N$, there exist a time $T > 0$ and a set of inputs u such that, if the system starts in state x_0 at time $t = 0$, it reaches x_f at time $t = T$. It should be noted that for nonlinear systems, the possibility of reaching from any given state a set of full dimension has also been called weak (local) controllability [8] and accessibility [18]. Here, following [17], we will refer to it as nonlinear reachability or simply reachability.

2.3.1. Linear Controllability and Reachability

A linear system given by (5) is reachable if and only if rank($\mathcal{C}$) $= n$, where $\mathcal{C}$ is the linear controllability matrix defined as:

$$\mathcal{C}(x) = \left(B|A \cdot B|A^2 \cdot B| \cdots |A^{n-1} \cdot B \right) \tag{8}$$

While the condition rank($\mathcal{C}$) $= n$ is both sufficient and necessary for reachability, for controllability it is only sufficient; that is, if A is singular, the system may be controllable even if rank($\mathcal{C}$) $< n$.

The "duality" between observability and controllability/reachability can be noticed by inspecting the linear observability (6) and controllability (8) matrices, since:

$$rank \left(B|A \cdot B|A^2 \cdot B| \cdots |A^{n-1} \cdot B \right) = n \Leftrightarrow$$
$$\Leftrightarrow rank \begin{pmatrix} B^T \\ B^T \cdot A^T \\ B^T \cdot (A^T)^2 \\ \vdots \\ B^T \cdot (A^T)^{n-1} \end{pmatrix} = n \tag{9}$$

Thus, the pair (A, B) is reachable if and only if the pair (A^T, B^T) is observable.

2.3.2. Nonlinear Controllability and Reachability

The nonlinear controllability matrix of the system (1) is as follows:

$$\mathcal{C}(x) = \left(\; f_2(x) \; \middle| \; L_{f_1} f_2(x) \; \middle| \; L_{f_1}^2 f_2(x) \; \middle| \; \cdots \; \middle| \; L_{f_1}^{n-1} f_2(x) \; \right) \tag{10}$$

The controllability matrix can be used to determine reachability with the following theorem:

Theorem 2. Controllability Rank Condition (CRC). *If the system M given by (1) satisfies rank($\mathcal{C}(x_0)$) $= n$, then M is (locally) reachable in a neighbourhood $N(x_0)$ of x_0 [17].*

Note that the CRC is a conservative criterion, since it provides only a sufficient—but not necessary—condition for reachability. A necessary condition can be obtained from the so-called controllability distribution $\Delta_c(x)$, which is computed iteratively as a sequence of distributions:

$$\Delta_0 = \text{span}\{f_2\} \;\rightarrow\; \Delta_1 \rightarrow \Delta_2 \rightarrow \ldots \rightarrow \Delta_{n-1} = \Delta_c, \tag{11}$$

with the following recursive rule:

$$\Delta_{i+1} = \text{span} \left\{ \Delta_i \bigcup \{[f_1, q], \; [f_2, q] : q(x) \in \Delta_i(x)\} \right\} \tag{12}$$

Theorem 3. Reachability Theorem. *The system M given by (1) is (locally) reachable around x_0 if and only if there exists a neighbourhood N of x_0 such that the distribution $\Delta_c(x)$, constructed as in (11,12), has constant dimension n for all $x \in N$ [17].*

While Theorem 3 provides a necessary and sufficient condition, it is more difficult to check than the CRC.

2.4. Structural Identifiability as Observability

Assuming that the model structure M (1) is correct, that the data is noise-free, and that the inputs to the system can be chosen freely, it is always possible to choose an estimated parameter vector $\hat{p}$

such that the model output $h(x, \hat{p})$ equals the one obtained with the true parameter vector, $h(x, p^*)$. If $\hat{p} = p^*$ this is trivially the case.

A parameter p_i is structurally locally identifiable if for almost any $p^* \in \mathbb{R}$ there is a neighbourhood $N(p^*)$ such that

$$\hat{p} \in N(p^*) \text{ and } h(x, \hat{p}) = h(x, p^*) \Rightarrow \hat{p}_i = p_i^* \tag{13}$$

A model M is structurally locally identifiable (s.l.i.) if all its parameters are s.l.i. If (13) does not hold in any neighbourhood of p^*, parameter p_i is structurally unidentifiable. A model M is structurally unidentifiable if at least one of its parameters is structurally unidentifiable.

Identifiability analysis can be formulated as a nonlinear observability problem [9,19]. To do so, let us augment the state variable vector so as to include also the model parameters:

$$\tilde{x} = \begin{bmatrix} x \\ p \end{bmatrix} \tag{14}$$

Accordingly, the f function that describes the time evolution of the state variables in equation (1) is augmented as follows:

$$\tilde{f}(\tilde{x}) = \begin{bmatrix} \dot{x} \\ \dot{p} \end{bmatrix} = \begin{bmatrix} f(x, u) \\ 0 \end{bmatrix} \tag{15}$$

where 0 is a zero-valued (since the parameters are constant in time) column vector of dimension $q \times 1$.

The resulting generalized observability-identifiability matrix, $O_I(\tilde{x})$, can be written as:

$$\mathcal{O}_I(\tilde{x}) = \begin{pmatrix} \frac{\partial}{\partial \tilde{x}} h(\tilde{x}) \\ \frac{\partial}{\partial \tilde{x}} (L_{\tilde{f}} h(\tilde{x})) \\ \frac{\partial}{\partial \tilde{x}} (L_{\tilde{f}^2} h(\tilde{x})) \\ \vdots \\ \frac{\partial}{\partial \tilde{x}} (L_{\tilde{f}^{n+q-1}} h(\tilde{x})) \end{pmatrix} \tag{16}$$

where $L_{\tilde{f}} h(\tilde{x})$ is a Lie derivative defined as in equation (2), that is, $L_{\tilde{f}} h(\tilde{x}) = \frac{\partial h(\tilde{x})}{\partial \tilde{x}} \tilde{f}(\tilde{x})$, and higher order derivatives are defined recursively as in equation (3).

Theorem 4. Observability-Identifiability Condition (OIC). *If M given by (1) satisfies $rank(\mathcal{O}_I(\tilde{x}_0)) = n + q$, then M is (locally) observable and identifiable in a neighbourhood $N(\tilde{x}_0)$ of $\tilde{x}_0$.*

Proof. It follows immediately from Theorem 1 and the definition of $\tilde{x}$. □

3. Results

Here we discuss the relationships between the structural properties defined in Section 2 and some issues that may appear. We illustrate them with several case studies, which are listed in Table 1 along with a summary of their properties.

3.1. Reachability Is Neither Necessary Nor Sufficient for Identifiability

For *linear* systems it was established early on that, despite the relationships that exist between those properties, observability and controllability are neither necessary nor sufficient conditions for structural identifiability. This was noted by DiStefano [20] by providing simple counter-examples (although DiStefano's brief note was disputed, see [21,22] and the author's replies to those comments). In a similar way, this section shows by means of three small examples that, for *nonlinear* systems, reachability—or lack thereof—has in principle no implications for structural identifiability.

Table 1. Main properties of the examples used in this paper. The term "generic" applied to SI (structural identifiability) or reachability means that the model has said property for almost all values of x. The last three columns refer to the possibility of losing said properties for certain values of x. N/A stands for Not Applicable.

Example	Ref.	(Generic) SI	(Generic) Reachability	Decrease in rank($\mathcal{O}_I$)	Loss of SI	Loss of Reachability
1	This paper	YES	NO	NO	NO	N/A
2	This paper	NO	YES	N/A	N/A	NO
3	[23]	NO	YES	N/A	N/A	YES
4	[11]	YES	NO	YES	YES	N/A
5	[24]	YES	NO	YES	NO	N/A
6.A	[14]	NO	NO	N/A	N/A	N/A
6.B	[14]	YES	NO	YES	NO	N/A

Example 1. *A model that is structurally identifiable but unreachable:*

$$M_1 : \begin{cases} \dot{x}_1 = p_1 \cdot (x_1^2 + A), \\ \dot{x}_2 = p_2 \cdot (x_1 \cdot x_2 + B) + u, \\ y_1 = x_1, \\ y_2 = x_2 \end{cases} \tag{17}$$

where $\{p_1, p_2\}$ are unknown parameters and $\{A, B\}$ known constants, all of which are positive quantities. The controllability distribution of M_1 is

$$\Delta_c(x) = \text{span} \left\{ \begin{array}{cc} 0 & 0 \\ 1 & -p_2 \cdot x_1 \end{array} \right\} \tag{18}$$

which has dimension $1 < n = 2$, so M_1 is unreachable (Theorem 3). Its observability-identifiability matrix is

$$\mathcal{O}_I(\tilde{x}) = \begin{pmatrix} 1 & 0 & 0 & 0 \\ 0 & 1 & 0 & 0 \\ 2 \cdot p_1 \cdot x_1 & 0 & x_1^2 + A & 0 \\ p_2 \cdot x_2 & p_2 \cdot x_1 & 0 & x_1 \cdot x_2 + B \end{pmatrix} \tag{19}$$

which has rank($\mathcal{O}_I(\tilde{x})$) $= 4 = n + q$, which means that it is observable and identifiable (OIC).

Example 2. *A model that is structurally unidentifiable but reachable:*

$$M_2 : \begin{cases} \dot{x}_1 = p_1 \cdot x_1 \cdot x_2^2 + u, \\ \dot{x}_2 = p_2 \cdot x_1, \\ y = p_3 \cdot x_1 \end{cases} \tag{20}$$

The controllability distribution of M_2 is

$$\Delta_c(x) = \text{span} \left\{ \begin{array}{cc} 1 & -p_1 \cdot x_2^2 \\ 0 & -p_2 \end{array} \right\} \tag{21}$$

which has dimension 2, so M_2 is reachable (Theorem 3). It is straightforward to calculate $\mathcal{O}_I(\tilde{x})$ (not shown here due to its large size) and see that rank($\mathcal{O}_I(\tilde{x})$) $= 3 < n + q = 5$, which means that it is unidentifiable (OIC). Specifically, p_2 is identifiable but p_1 and p_3 are not. It can also be noticed that, according to the classic definition of observability (which assumes that the parameter values are known) model M_2 is observable, since

$$\mathcal{O}(x) = \begin{pmatrix} 1 & 0 \\ p_1 \cdot x_2^2 & 2 \cdot p_1 \cdot x_1 \cdot x_2 \end{pmatrix} \tag{22}$$

and thus $\mathrm{rank}(\mathcal{O}(x)) = 2$ and the ORC holds.

Example 3. *A model that is structurally unidentifiable but (almost everywhere) reachable:*

$$M_3 : \begin{cases} \dot{x}_1 &= p_1 \cdot x_1 \cdot x_2, \\ \dot{x}_2 &= p_2 \cdot u, \\ y &= x_1 \end{cases} \tag{23}$$

This model is taken from [23]. For this system $n + q = 4$ and $\mathrm{rank}(O_I(\tilde{x})) = 3$. Thus the observability-identifiability condition (OIC) of Theorem 4 does not hold; the model as a whole, and $\{p_1, p_2\}$ in particular, are structurally locally unidentifiable. This unidentifiability result was obtained symbolically, for generic values of the initial conditions. To illustrate this lack of identifiability, the time evolution of M_3 is shown in Figure 1 for two different parameter vectors: $\{p_1 = 1, p_2 = 1\}$ and $\{p_1 = 1.6, p_2 = 1.25\}$. If the model is simulated from the initial conditions $\{x_1(0), x_2(0)\} = \{1, 1\}$, the model outputs are distinguishable for the two different parameter vectors (panel A). However, when the model is simulated from $\{x_1(0), x_2(0)\} = \{1, 0\}$, the outputs are identical (panel B): for these initial conditions the parameters are indistinguishable, i.e., the model is unidentifiable. Furthermore, even if the model is started from a different initial condition, $x_2 \neq 0$, its output may be the same to that of $x_2 = 0$ for two different parameter vectors (panel C). Thus, the structural unidentifiability of this model is not exclusive of the initial condition $x_2 = 0$. Note that, however, if we were able to measure not only x_1 but also x_2, the model would become identifiable, since the time course of x_2 is different in each of the aforementioned cases.

Regarding reachability, the controllability distribution of M_3 is

$$\Delta_c(x) = \mathrm{span} \left\{ \begin{array}{cc} 0 & -p_1 \cdot p_2 \cdot x_1 \\ p_2 & 0 \end{array} \right\} \tag{24}$$

which has dimension 2, so M_3 is reachable (Theorem 3). Thus, M_3 is generically reachable and structurally unidentifiable. So is M_2; however, there is a difference between both models: unlike in the case of M_2, the dimension of the controllability distribution of M_3 can decrease for certain values of x: specifically, for $x_1 = 0$, its dimension decreases from 2 to 1, as can be seen in Equation (24), and in that particular case the model is no longer reachable. This is illustrated in panel D of Figure 1, which shows that if $x_1(0) = 0 \Rightarrow x_1(t) = 0 \ \forall t$. We remark that this loss of reachability is not the cause of the model's unidentifiability, since the model is also unidentifiable for the initial conditions shown in the other panels, for which there is no loss of reachability.

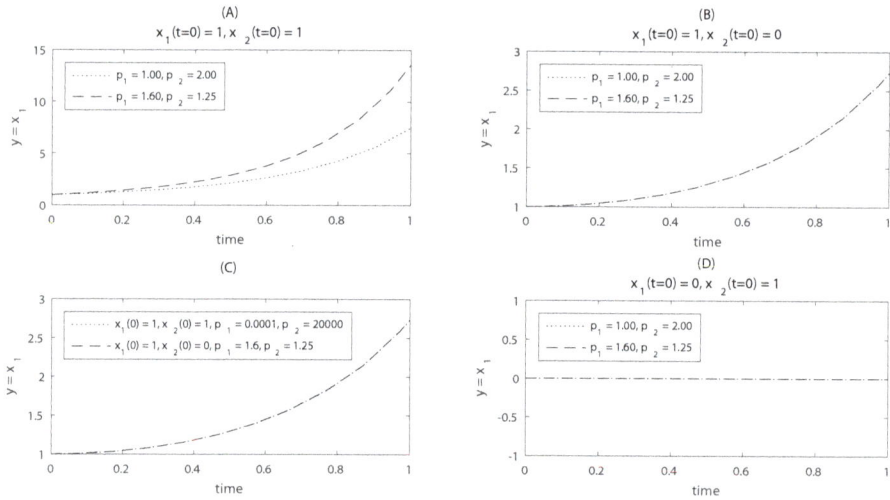

Figure 1. Time courses of the M_3 model used in Example 3. Panel (**A**) shows the evolution of the system starting from non-zero initial conditions, for two different parameter vectors. From this plot it would seem that the model is structurally identifiable, since different parameter vectors yield different model outputs; Panel (**B**) shows the same time courses for initial condition $x_2(t = 0) = 0$; in this case the model output is identical for two different parameter vectors, which makes the parameters indistinguishable; Panel (**C**) shows that it is actually impossible to distinguish between initial condition $x_2(t = 0) = 0$ and initial condition $x_2(t = 0) = 1$: the output of the model simulated with two wildly different parameter vectors can be the same. This illustrates that unidentifiability is inescapable if we only measure x_1; Panel (**D**) shows the model output for two parameter vectors and initial condition $x_1(t = 0) = 0$; for this case the model output remains at zero, and there is a loss of reachability.

3.2. The Results of Identifiability Tests Can Be Misleading for Certain State Vectors

The OIC of Theorem 4 represents a general result, which is valid for all values of the states and parameters *except for a set of measure zero*. For these exceptions, however, its results may be misleading. This section illustrates this fact with an example.

Example 4. *A structurally identifiable model that loses identifiability for certain values of x:*
Consider the following biochemical network [11]:

$$E + S \underset{p_2}{\overset{p_0}{\rightleftharpoons}} ES \xrightarrow{p_3} E + P; \; P \xrightarrow{p_1} \varnothing \tag{25}$$

If we denote by x_1, x_2, and x_3 the concentrations of substrate (S), enzyme (E), and product (P), respectively, this system can be modelled by the following equations [11]:

$$M_4 : \begin{cases} \dot{x}_1 = -x_1 \cdot x_2 + p_2 \cdot (10 - x_2), \\ \dot{x}_2 = -x_1 \cdot x_2 + (p_2 + p_3) \cdot (10 - x_2), \\ \dot{x}_3 = -p_1 \cdot x_3 + p_3 \cdot (10 - x_2) \\ y_1 = x_1, y_2 = x_3 \end{cases} \tag{26}$$

Note that these equations were obtained by August and Papachristodoulou [11] after making two assumptions: (i) that $[ES] + [E] = 10$, which allows omitting [ES] from the equations by replacing it with $[ES] = 10 - [E] = 10 - x_2$; and (ii) that the reaction rates are normalized, such that $p_0 = 1$ is a known constant.

For this example $n + q = 6$ and $\text{rank}(O_I(\tilde{x})) = 6$, which means that the OIC of Theorem 4 holds. Therefore methods based in the OIC [10,11,13] classify the model as observable and identifiable for generic values of the states and parameters. However, note that for the particular initial conditions $\{x_1(0) = 0,\ x_2(0) = 10\}$, two of the states remain at zero $\{x_1(t) = 0,\ x_2(t) = 0\}\ \forall t \geq 0$, and in this case only p_1 appears in the equations. Thus, *for these particular initial conditions the model is not identifiable*, since it is not possible to determine the values of p_2 and p_3. Obviously, model M_4 is not reachable, since it does not have any control inputs and thus its controllability distribution is empty. Therefore the loss of identifiability cannot be attributed to a loss of reachability for certain initial conditions. However, it is linked to the fact that $\{x_1,\ x_2\}$ remain equal to zero if the system is started from $\{x_1(0) = 0,\ x_2(0) = 10\}$. This situation is not rare: indeed, by examining the Equation (17) that describe model M_1, it can be noticed that if $A = 0$, state x_1 would remain equal to zero if started in that condition, and the model would be unidentifiable, since the third column of its observability-identifiability matrix (19) would be zero.

Example 4 shows how a model that is "generally" structurally identifiable—that is, it is identifiable for *almost* all values of its state variables—can lose its identifiability for certain particular values of its states (or equivalently, for certain initial conditions). This loss of identifiability results from some of the states of M_4 remaining at zero value from certain initial conditions, which prevents some parameters appearing in the equations of those states from being identified. This loss of identifiability causes a loss of rank in the OIC: if the model states are replaced with the initial conditions of interest, x_0*, it results that $\text{rank}(O_I(x_0*)) < n + q$; thus the loss of identifiability entails a rank deficiency.

However, this rank check is not a definitive proof: the next model provides a counter-example that shows that, even when the rank of a (generally structurally identifiable) matrix decreases for certain initial conditions, this does not necessarily result in structural unidentifiability.

Example 5. *A structurally identifiable model despite rank deficiency of O_I for certain values of x:*
Robertson [24] proposed as a case study an autocatalytic reaction with the following scheme:

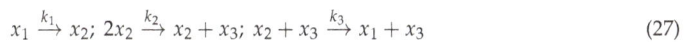

$$x_1 \xrightarrow{k_1} x_2;\ 2x_2 \xrightarrow{k_2} x_2 + x_3;\ x_2 + x_3 \xrightarrow{k_3} x_1 + x_3 \tag{27}$$

Its kinetics are given by the following equations:

$$M_5 : \begin{cases} \dot{x}_1 = -k_1 \cdot x_1 + k_3 \cdot x_2 \cdot x_3, \\ \dot{x}_2 = k_1 \cdot x_1 - k_2 \cdot x_2^2 - k_3 \cdot x_2 \cdot x_3, \\ \dot{x}_3 = k_2 \cdot x_2^2, \\ y_1 = x_1, y_2 = x_2 \end{cases} \tag{28}$$

This model was showcased in [25] as an example of an unidentifiable model, and later used as a case study in [26]. Specifically, Eydgahi et al. stated that "the inability of estimation to recover the parameter values used to generate synthetic data is not due to problems with the computational procedures. Instead, it represents *a fundamental limit* on our ability to understand biochemical systems based on time-course data alone". However, this model is in fact structurally identifiable (O_I is full rank for generic values of x), although its *practical* identifiability is poor. A reason for its lack of practical identifiability is that, for the nominal parameter values, x_2 has very low values compared to the other two states (six orders of magnitude smaller). Thus, the difficulty in recovering the true values of its parameters is not due to a *structural* deficiency, but to *numerical* limitations.

Interestingly, when $x_2(0) = x_3(0) = 0$, the observability-identifiability matrix O_I is not full rank, so the system may seem to be structurally unidentifiable from those initial conditions. However, such a loss of identifiability is "strictly local". By this we mean that, as can be seen in Figure 2, all the states (including x_2 and x_3) reach non-zero values immediately after $t = 0$ (note that the concentration curves have been normalized to adopt values between 0 and 1, to facilitate the visualization). For example, at $t = 10$ the state vector is $\{x_1(10) = 0.8414, x_2(10) = 1.623 \cdot 10^{-5}, x_3(10) = 0.1586\}$, and O_I has full rank when evaluated at that point. Therefore the model is still identifiable when simulated from these initial conditions. This example shows that, even when the O_I loses its full rank for certain values of the state variables, this does not necessarily lead to loss of identifiability, as long as it does not produce loss of accessibility.

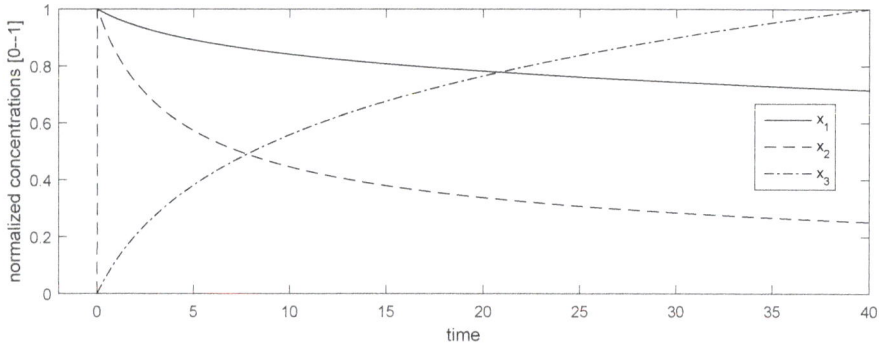

Figure 2. Time courses of the Robertson model used in Example 5. The vertical axis shows the concentration curves, which have been normalized to adopt values between 0 and 1 to facilitate the visualization.

3.3. Knowledge of Additional Initial Conditions Can Increase Identifiability

So far we have assumed that the initial conditions of the states corresponding to measured outputs are known, and that those of unmeasured states are unknown. This is the typical situation: in general we do not have information about unmeasured states; on the other hand, since structural identifiability analysis assumes unlimited measurements, we can consider all the values of the measured states known—including, naturally, their initial conditions.

However, in certain cases we may possess information about the initial condition of a state, even if we are not able to measure its subsequent behaviour. In this case, such additional information can help identify some parameters that would be unidentifiable without it. This possibility was noted by Chappell and Godfrey [27], who used a case study which they claimed to be the first example of a real-life nonlinear model that was not globally identifiable. Here we illustrate this case with the following example:

Example 6. *A structurally unidentifiable model that becomes identifiable if the initial conditions (of unobserved states) are known:*

Denis-Vidal et al [14] proposed the following model to illustrate how the identifiability of uncontrolled models may depend on the initial conditions.

$$M_6 : \begin{cases} \dot{x}_1 = \theta_1 \cdot x_1^2 + \theta_2 \cdot x_1 \cdot x_2, \ x_1(0) = 1 \\ \dot{x}_2 = \theta_3 \cdot x_1^2 + x_1 \cdot x_2, \ x_2(0) = b \\ y = x_1 \end{cases} \tag{29}$$

It can be seen that parameters θ_2 and θ_3 are structurally unidentifiable; we refer to this case as Example 6.A (as written in Table 1). However, if we know the initial condition of the unmeasured state, $x_2(0) = b$, then the model becomes structurally identifiable; we refer to this variant as Example 6.B. In this case, a decrease in the rank of O_I (from full rank, 5, to 4) takes place if $b = 0$, but, similarly to Example 5, the state x_2 immediately departs from zero and the model is structurally identifiable.

3.4. Guaranteeing the Results of the Structural Identifiability Test

The examples of the preceding section show that local structural identifiability is a *generic* property that may be lost for *particular* values of the state variables, and that such loss of identifiability may not be detected by differential geometry methods such as those based on the OIC. This issue is shared by differential algebra methods: for them, [15,28] concluded that the result of the identifiability test is still correct if a model is *globally* inaccessible (unreachable), but problems appear when a model is only inaccessible from initial points belonging to a thin set.

The realization in [15,28] suggests performing an additional check on the reachability of a system: if for a certain state x_0 we find both a decrease in the rank of the observability-identifiability matrix $O_I(x_0)$ *and* a decrease in the dimension of the controllability distribution $\Delta_c(x_0)$, we can conclude that the model is not identifiable when started from the initial conditions x_0. However, there are issues that cannot be revealed by this test: we have seen that such a loss of identifiability for particular x_0 can also happen in uncontrolled models such as Example 4, which are by definition unreachable and for which the dimension of Δ_c is always zero. Therefore, for such cases an alternative is to simulate the model from x_0—as we did for Example 5 in Figure 2—to see whether there are states that remain at zero. In general, structural identifiability can be assessed for particular initial conditions x_0 as follows:

1. Check the Observability-Identifiability condition (OIC, Theorem 4) for a *generic symbolic* augmented vector $\tilde{x}$. If the matrix is *not* full rank, i.e., rank$(\mathcal{O}_I(\tilde{x})) < n + q$, then the model is structurally unidentifiable and no further tests are needed. If, however, it is full rank, i.e., rank$(\mathcal{O}_I(\tilde{x})) = n + q$, the model is generically structurally identifiable. However, in this case there may be loss of identifiability for certain states; to assess this we proceed to the next step.
2. Check the Observability-Identifiability condition (OIC, Theorem 4) for a *particular* vector of initial conditions of interest, $\tilde{x}_0$. If rank$(\mathcal{O}_I(\tilde{x}_0)) < n + q$, there has been a loss of rank which indicates loss of local identifiability (as in Example 4), but that may be possible to overcome (as in Example 5). To assess this point, we go to step 3.
3. Simulate the model starting from x_0 to see if there are any states that remain zero.

In the present work we have carried the rank calculations using a recently presented tool called STRIKE-GOLDD [13]. STRIKE-GOLDD is a methodology for local structural identifiability analysis that evaluates the OIC symbolically. Although it considers in principle generic initial conditions, it can also test particular initial conditions by specializing the generic x in the O_I matrix to the particular initial conditions of interest, x_0. If the system under study is rational, it is also possible to carry out a similar test with the Exact Arithmetic Rank (EAR) method [10]. EAR allows specifying initial conditions and uses a computationally efficient numerical procedure for obtaining rank$(O_I(x_0))$. Unlike EAR, STRIKE-GOLDD evaluates the OIC symbolically and does not require that the system is rational, although this generality results in a computationally more expensive procedure. In another related work, [11] evaluated the OIC using a more conservative and computationally expensive procedure, recasting the rank calculation task as a sum-of-squares optimization problem. Its advantage is that it provides a result that is not only valid for a particular point x_0 but for range of values, that is, for all states that fall within certain bounds, $x_L < x < x_U$. Unfortunately, these calculations are only feasible for small systems due to computational limitations. For rational systems it is also possible to use the differential algebra method DAISY [29], which allows specifying initial conditions. It is important to be aware of the advantages and limitations of these methods when choosing one for assessing the structural identifiability of a model.

3.5. Finding Specific Initial Conditions That Lead to Loss of Rank in the OIC

If the initial conditions are not fixed *a priori*, a question naturally arises: given a nonlinear model that is generally structurally identifiable, is it possible to find the specific values of the initial conditions—if there exist any—for which the OIC does not hold?

3.5.1. Finding Solutions Analytically

It may be possible to find such values analytically. For example, by examining Equation (26) it can be observed that for $\{x_1 = 0, x_2 = 10, x_3 = 0\}$ the right hand side of the ODEs are made zero and the system remains in a steady state, thus eliminating any influence of the parameters on the output and rendering them unidentifiable. However, there are also other combinations that lead to loss of identifiability. A general way of finding these initial conditions is to calculate the singular values of O_I using a symbolic software, such as Mathematica or the MATLAB Symbolic Math Toolbox. Then, the state vectors x^* that decrease $\text{rank}(O_I(x^*))$ can be found by equating the expressions of the singular values to zero and solving the resulting equations. However, this approach involves symbolic calculation of the singular values, which is a very complex task for which explicit solutions can be obtained only for very small models. For example, this approach did not yield results for a model of moderate size such as the one in Example 4, even after fixing the parameter values to random numbers to reduce the problem size.

3.5.2. Finding Solutions via Optimization

An alternative, generally applicable way of answering this question is by formulating it as an optimization problem. In it, the decision variables are the system states, and the objective to minimize is the rank of the observability-identifiability matrix. Is this rank can be made zero for particular values of the state variables, those values correspond to initial conditions that lead to lack of structural identifiability. Since the rank is an integer, it is more convenient for computational optimization to use as the objective function the smallest singular value of O_I, which is a positive real number and thus leads to a continuous function value. Thus it is possible to calculate a singular value decomposition of O_I numerically and use its smallest singular value as the objective; if it can be made zero, the matrix is not full rank and structural identifiability is lost. In order to calculate a numeric value for the singular values we must replace the symbolic variables by numbers; using non-repeated prime numbers minimizes the risk of accidental cancellations that could artificially reduce the rank. Then the optimization problem can be mathematically formulated as follows:

$$\min_{x_L < x < x_U} \{f_{obj} = \inf[\text{sing}(\mathcal{O}_I(\tilde{x}))]\} \tag{30}$$

That is: find the vector x that minimizes the objective function consisting of the smallest singular value of the observability-identifiability matrix, for values of the states that lie within some lower and upper bounds, $x_L < x < x_U$.

We applied this strategy to example 4, using as optimization method a general purpose metaheuristic called enhanced scatter search (eSS), from the MATLAB (version R2015b, MathWorks, Natick, MA, USA) implementation of the MEIGO toolbox [30]. Setting $\{x_L = 0, x_U = 20\}$ as bounds for the decision variables, and using FMINCON as local search method, the algorithm reported $f_{obj} = 0$ after a few minutes, with the optimal decision variables being $[x_1 = 0, x_2 = 10, x_3 = 6.9081]$. Since the optimization method is not deterministic, different outcomes can be obtained. Running the algorithm again we encountered the solution $[x_1 = 0, x_2 = 10, x_3 = 3.9086]$; in this way it is easy to realize that any vector in which $[x_1 = 0, x_2 = 10]$ leads to unidentifiability regardless of the value of x_3.

4. Conclusions

Structural properties (those that are determined exclusively by the model equations) provide information about the behaviour of a system. Their analysis can reveal the existence of limitations regarding system dynamics and model identification. In this paper we dealt with the relations between the systemic properties of observability, reachability, and identifiability of nonlinear models. Identifiability and observability are tightly related, since structural identifiability can be recast as a "generalized" observability property (also called "augmented" or "extended"). On the other hand, observability and reachability (or similar terms like controllability and accessibility) are usually considered as dual concepts.

Reachability is not a requisite for observability nor identifiability, nor vice versa. However, in some cases a system which is in principle identifiable can become unidentifiable due to some of its states becoming inaccessible (unreachable) from certain initial conditions. This problem is not always detected by structural identifiability analysis methods. In general, the results of such methods—for example, those based on differential geometry and differential algebra approaches—are valid for "almost all" combinations of state and parameter values (i.e., for a dense subset of the state-parameter space), but there may be exceptions for values belonging to thin sets (i.e., isolated values). To investigate this possibility we propose to calculate the rank of the generalized observability-identifiability matrix (O_I) not only symbolically (which provides the generic result), but also numerically for the particular initial conditions of interest. We have noted however that a loss of rank in the O_I for particular initial conditions x_0 does not always imply a loss of structural identifiability: in some cases, such as Example 5, the model can still be identified because the time evolution of the system escapes from the pathological state. Thus, to assess if there is a loss of identifiability it is advisable to simulate the model for the initial conditions of interest.

We have also noted in this manuscript that loss of reachability is not the only possible cause of a loss of identifiability for specific initial conditions. Indeed, such unidentifiability can arise also in uncontrolled models, whose controllability distribution Δ_C is zero, as seen in Example 4. For this reason we have seen that it is not entirely appropriate to assess the loss of identifiability by looking for a possible decrease in the dimension of Δ_C, and the aforementioned simulation-based analysis should be preferred instead.

The related problem of finding which initial conditions lead to loss of identifiability is in general more complicated. When such initial conditions are difficult to determine analytically, an alternative is to use an optimization procedure. We have demonstrated the feasibility of this approach using an example taken from the literature [11], for which this loss of identifiability had not been reported before.

We would like to mention that the idea of optimizing initial conditions to achieve a goal related with identifiability has been applied in the literature before, albeit with a totally different purpose: in the context of optimal experiment design, it may be desirable to find initial conditions that *maximize* the *practical* identifiability of the parameters, as done e.g., by [31]. In contrast, here we have used it for what can be considered as the opposite task: finding initial conditions that *minimize* (destroy) *structural* identifiability. Both calculations provide useful information in different contexts.

The considerations presented in this paper should be taken into account before performing tasks such as design of experiments or parameter estimation, in order to prevent identifiability issues that may potentially render their results useless. For example, any attempts at calibrating a structurally unidentifiable model will fail, resulting in a waste of time and effort and in wrong parameter estimates. Furthermore, if this structural unidentifiability is mistaken for *practical* unidentifiability (a related but different problem), it may lead to trying to solve the problem by investing additional efforts in designing and performing new experiments, which will nevertheless be sterile. Therefore it is advisable to assess the structural identifiability of a model for all the particular conditions of interest before attempting at calibrating and further exploiting it.

Acknowledgments: This work has received funding from the European Union's Horizon 2020 research and innovation programme under grant agreement No 686282 ("CANPATHPRO").

Author Contributions: A.F.V. conceived of the study; A.F.V. and J.R.B. designed the computational experiments; A.F.V. performed the computations; A.F.V. and J.R.B. analyzed the results; A.F.V. and J.R.B. wrote the paper.

Conflicts of Interest: The authors declare no conflict of interest. The founding sponsors had no role in the design of the study; in the collection, analyses, or interpretation of data; in the writing of the manuscript, and in the decision to publish the results.

Abbreviations

The following abbreviations are used in this manuscript:

ORC   Observability Rank Condition
CRC   Controllability Rank Condition
OIC   Observability-Identifiability Condition
ODE   Ordinary Differential Equation
EAR   Exact Arithmetic Rank
SI   Structural Identifiability

References

1. Walter, E.; Pronzato, L. *Identification of Parametric Models From Experimental Data*; Communications and Control Engineering Series; Springer: London, UK, 1997.
2. DiStefano J., III. *Dynamic Systems Biology Modeling and Simulation*; Academic Press: New York, NY, USA, 2015.
3. Villaverde, A.F.; Barreiro, A. Identifiability of large nonlinear biochemical networks. *MATCH Commun. Math. Comput. Chem.* **2016**, *76*, 259–276.
4. Bellman, R.; Åström, K.J. On structural identifiability. *Math. Biosci.* **1970**, *7*, 329–339.
5. Chiş, O.T.; Banga, J.R.; Balsa-Canto, E. Structural identifiability of systems biology models: A critical comparison of methods. *PLoS ONE* **2011**, *6*, e27755.
6. Miao, H.; Xia, X.; Perelson, A.S.; Wu, H. On identifiability of nonlinear ODE models and applications in viral dynamics. *SIAM Rev.* **2011**, *53*, 3–39.
7. Kalman, R.E. Contributions to the theory of optimal control. *Bol. Soc. Mat. Mexicana* **1960**, *5*, 102–119.
8. Hermann, R.; Krener, A.J. Nonlinear controllability and observability. *IEEE Trans. Autom. Control* **1977**, *22*, 728–740.
9. Tunali, E.T.; Tarn, T.J. New results for identifiability of nonlinear systems. *IEEE Trans. Autom. Control* **1987**, *32*, 146–154.
10. Karlsson, J.; Anguelova, M.; Jirstrand, M. An Efficient Method for Structural Identiability Analysis of Large Dynamic Systems. In Proceedings of the 16th IFAC Symposium on System Identification, Brussels, Belgium, 11–13 July 2012; Volume 16, pp. 941–946.
11. August, E.; Papachristodoulou, A. A new computational tool for establishing model parameter identifiability. *J. Comput. Biol.* **2009**, *16*, 875–885.
12. Chatzis, M.N.; Chatzi, E.N.; Smyth, A.W. On the observability and identifiability of nonlinear structural and mechanical systems. *Struct. Control Health Monit.* **2015**, *22*, 574–593.
13. Villaverde, A.F.; Barreiro, A.; Papachristodoulou, A. Structural Identifiability of Dynamic Systems Biology Models. *PLOS Comput. Biol.* **2016**, *12*, e1005153.
14. Denis-Vidal, L.; Joly-Blanchard, G.; Noiret, C. Some effective approaches to check the identifiability of uncontrolled nonlinear systems. *Math. Comput. Simul* **2001**, *57*, 35–44.
15. Saccomani, M.P.; Audoly, S.; D'Angiò, L. Parameter identifiability of nonlinear systems: The role of initial conditions. *Automatica* **2003**, *39*, 619–632.
16. Xia, X.; Moog, C.H. Identifiability of nonlinear systems with application to HIV/AIDS models. *IEEE Trans. Autom. Control* **2003**, *48*, 330–336.
17. Vidyasagar, M. *Nonlinear Systems Analysis*; Prentice Hall: Englewood Cliffs, NJ, USA, 1993.
18. Sontag, E.D. *Mathematical Control Theory: Deterministic Finite Dimensional Systems*; Springer Science & Business Media: Berlin, Germany, 2013; Volume 6.
19. Sedoglavic, A. A probabilistic algorithm to test local algebraic observability in polynomial time. *J. Symb. Comput.* **2002**, *33*, 735–755.
20. DiStefano J., III. On the relationships between structural identifiability and the controllability, observability properties. *IEEE Trans. Autom. Control* **1977**, *22*, 652–652.
21. Cobelli, C.; Lepschy, A.; Romanin-Jacur, G. Comments on "On the relationships between structural identifiability and the controllability, observability properties". *IEEE Trans. Autom. Control* **1978**, *23*, 965–966.

22. Jacquez, J. Further comments on "On the relationships between structural identifiability and the controllability, observability properties". *IEEE Trans. Autom. Control* **1978**, *23*, 966–967.

23. Balsa-Canto, E. Tutorial on Advanced Model Identification using Global Optimization. In Proceedings of the ICSB 2010 International Conference on Systems Biology, Edinburgh, UK, 11–14 October 2010.

24. Robertson, H. The solution of a set of reaction rate equations. In *Numerical Analysis: An Introduction*; Academic Press: London, UK, 1966; pp. 178–182.

25. Eydgahi, H.; Chen, W.W.; Muhlich, J.L.; Vitkup, D.; Tsitsiklis, J.N.; Sorger, P.K. Properties of cell death models calibrated and compared using Bayesian approaches. *Mol. Syst. Biol.* **2013**, *9*, 644.

26. Mannakee, B.K.; Ragsdale, A.P.; Transtrum, M.K.; Gutenkunst, R.N. Sloppiness and the geometry of parameter space. In *Uncertainty in Biology*; Springer: Cham, Switzerland, 2016; pp. 271–299.

27. Chappell, M.J.; Godfrey, K.R. Structural identifiability of the parameters of a nonlinear batch reactor model. *Math. Biosci.* **1992**, *108*, 241–251.

28. D'Angiò, L.; Saccomani, M.P.; Audoly, S.; Bellu, G. Identifiability of Nonaccessible Nonlinear Systems. In *Positive Systems*; Bru, R., Romero-Vivó., Eds.; Springer: Basel, Switzerland, 2009; Volume 389, *LNCIS*, pp. 269–277.

29. Bellu, G.; Saccomani, M.P.; Audoly, S.; D´Angio, L. DAISY: A new software tool to test global identifiability of biological and physiological systems. *Comput. Methods Programs Biomed.* **2007**, *88*, 52–61.

30. Egea, J.A.; Henriques, D.; Cokelaer, T.; Villaverde, A.F.; MacNamara, A.; Danciu, D.P.; Banga, J.R.; Saez-Rodriguez, J. MEIGO: An open-source software suite based on metaheuristics for global optimization in systems biology and bioinformatics. *BMC Bioinf.* **2014**, *15*, 136.

31. Balsa-Canto, E.; Alonso, A.; Banga, J.R. An iterative identification procedure for dynamic modeling of biochemical networks. *BMC Syst. Biol.* **2010**, *4*, 11.

Article

Optimal Experimental Design for Parameter Estimation of an IL-6 Signaling Model

Andrew Sinkoe [1,2] **and Juergen Hahn** [1,2,3,*]

1   Department of Biomedical Engineering, Rensselaer Polytechnic Institute, Troy, NY 12180, USA
2   Center for Biotechnology and Interdisciplinary Studies, Rensselaer Polytechnic Institute,
    Troy, NY 12180, USA
3   Department of Chemical & Biological Engineering, Rensselaer Polytechnic Institute, Troy, NY 12180, USA
*   Correspondence: hahnj@rpi.edu; Tel.: +1-518-276-2138

Received: 16 June 2017; Accepted: 22 August 2017; Published: 1 September 2017

Abstract: IL-6 signaling plays an important role in inflammatory processes in the body. While a number of models for IL-6 signaling are available, the parameters associated with these models vary from case to case as they are non-trivial to determine. In this study, optimal experimental design is utilized to reduce the parameter uncertainty of an IL-6 signaling model consisting of ordinary differential equations, thereby increasing the accuracy of the estimated parameter values and, potentially, the model itself. The D-optimality criterion, operating on the Fisher information matrix and, separately, on a sensitivity matrix computed from the Morris method, was used as the objective function for the optimal experimental design problem. Optimal input functions for model parameter estimation were identified by solving the optimal experimental design problem, and the resulting input functions were shown to significantly decrease parameter uncertainty in simulated experiments. Interestingly, the determined optimal input functions took on the shape of PRBS signals even though there were no restrictions on their nature. Future work should corroborate these findings by applying the determined optimal experimental design on a real experiment.

Keywords: optimal experimental design; D-optimality criterion; Fisher information matrix; sensitivity analysis; IL-6 signaling; parameter estimation; piecewise constant functions

---

1. Introduction

Mathematical models of intracellular signaling pathways are important for understanding and predicting how cells respond to certain stimuli. Such models can be modified readily when new findings become available, and can be a useful tool in directing new studies based on hypotheses generated by the model's predictions.

Interleukin-6 (IL-6) is a cytokine involved in a variety of inflammatory processes [1–4]. Understanding the signaling pathways associated with extracellular IL-6 excitation is important for elucidating and modulating the biological response to inflammation. Because chronic inflammation can cause tissue damage and poses a serious health risk during chronic infection or autoimmune conditions [5–7], therapeutic treatments for chronic inflammation are an active area of research. These efforts can be augmented through the use of mathematical models for inflammatory signaling, where IL-6 plays a major role.

An IL-6 signaling model (Appendix A) was previously developed in [1–4]. Originally, a detailed model containing 77 ordinary differential equations (ODEs) and 128 parameters was derived, followed by a model simplification procedure to decrease the number of ODEs and parameters to 13 and 19, respectively. The simplified model contains only the variables and parameters deemed necessary for representing the dynamics of the signaling system, as determined by parameter sensitivity analysis and observability analysis [2]. This simplified model is the subject of the present study, and will henceforth be referred to as the 'IL-6 model' or the 'model' while the original detailed model will be referred to as the 'original IL-6 model' or the 'original model'.

In this study, optimal experimental design was applied to increase the accuracy of the 19 parameters of the IL-6 model. The parameters in this model correspond to rate constants for the chemical reactions in the signaling pathway represented by the model. By increasing the accuracy—and conversely reducing the uncertainty—of the model parameters, it is possible to obtain a more accurate model through iterative model adjustment. To demonstrate and examine this step in the model development process, optimal experimental design methods were applied in this study to minimize the uncertainty of the IL-6 model parameters estimated from least squares with experimental data.

Optimal experimental design has been utilized for decades in a variety of settings in which it is of interest to maximize efficiency of resource use and obtain a significant amount of information from experiments with acceptable cost [8–16]. Recently, as biological modeling and systems biology have emerged as an important area in biomedical research, optimal experimental design applied to biological experimental systems has become more popular [17–28]; additionally, optimal experimental design has been recognized as a valuable tool in optimal control for several decades [29]. For example, Jones et al. [13] maximized production of an exogenous commodity chemical in metabolically engineered *E. coli* using an empirical modeling method similar to those used in [15,16] to maximize the efficacy of drug delivery. Weber [26] utilized optimal experimental design to maximize model prediction accuracy for a model of vesicle transport via the *trans*-Golgi network. Bandara [18] performed optimal experimental design to reduce parameter uncertainty in a model of phosphatidylinositol 3,4,5-trisphosphate signaling. These studies demonstrate the effectiveness of optimal experimental design for obtaining maximally informative experimental data.

Here, optimal experimental design is applied to the problem of maximizing parameter accuracy for the IL-6 model. In particular, the D-optimality criterion, applied to the Fisher information matrix (FIM), is maximized over a set of IL-6 concentration input functions to determine an optimal dynamic IL-6 input profile for exciting the signaling system to generate data for least squares estimation of the model parameters [30]. The experimental design constraints considered in the problem are based on available resources and limitations present in a typical laboratory capable of performing the designed experiments. Namely, it is assumed that (a) the measurements of protein concentrations for two transcription factors are recorded as a time series; (b) the sampling time between measurements does not change during the course of an experiment; (c) the model represents signal transduction in rat hepatocytes which are stimulated in vitro with IL-6 following a dynamic input function; and (d) the IL-6 concentration is kept below cytotoxic levels. A piecewise constant IL-6 input function was computed by solving the optimal experimental design problem with the D-optimality criterion, operating on the Fisher information matrix, as the objective function. Since the Fisher information matrix contains only local information, a sensitivity matrix was also computed using the Morris method and the D-optimality criterion was applied to the sensitivity matrix as well in order to corroborate that the results are not just local in nature. The optimal IL-6 input function was found to substantially decrease the model parameter uncertainty in simulated least squares fits compared to a constant stimulation with IL-6.

The paper is organized as follows: optimal experimental design and its application to ODE models are presented in Section 2; application of optimal experimental design to the specific case of the IL-6 model is presented in Section 3 and results from solving the optimal experimental design problem are presented in Section 4, including simulation results for calculating parameter uncertainty; Section 5 discusses implications and suggests further applications of the optimal experimental design methodology presented here. The IL-6 model is included for reference in Appendix A.

2. Optimal Experimental Design

An ordinary differential equation model, e.g., for describing signal transduction, can be written as

$$\frac{dx}{dt} = f(t, x, u, p); \; y = g(t, x, u, p), \tag{1}$$

where x is a time-dependent vector of state variables, u is a time-dependent controlled input to the system, p is a vector of constant parameters, and y is a vector of measured quantities related to the model variables. Often, as is the case here, y is simply a subset of state variables from the vector x that are measured experimentally over time. While these measurements y are discrete in nature, the assumption that y is a continuous variable can be made if the sampling frequency is sufficiently high.

Optimal experimental design involves maximizing a criterion function that indicates the quantity of information gained by a given experiment, often in the context of model identification [8–12,14,17–23,29,31]. Several commonly used criterion functions for experimental design exist. One of these is A-optimality, which seeks to minimize the trace of the inverse of the Fisher information matrix. This criterion results in minimizing the average variance of the estimates of the regression coefficients. Another popular approach is E-optimality which maximizes the smallest eigenvalue of the Fisher information matrix. However, the most popular approach is D-optimal experimental design, i.e., which maximizes the determinant of the FIM [14,19,24]. The D-optimality criterion was chosen for this work as it seeks to minimize the covariance of the parameter estimates for a specified model [14]. It should be noted that no experimental design is optimal in all aspects and maximizing one particular criterion often negatively affects the experimental design of other criteria. That being said, the D-optimality criterion has found widespread use in practice as it results in experimental designs that have many of the properties that one usually looks for. For numerical considerations, the natural logarithm of the determinant of the FIM was taken for the problem presented. The optimal experimental design objective can be written generally as

$$\max \varphi_D(\mathbf{F}), \tag{2}$$

where $\mathbf{F}$ is the FIM, defined in more detail below, and φ_D is the D-optimality criterion.

Taking the determinant of the parameter covariance matrix corresponds to computing the volume of the parameter space that would allow for a solution to the least squares parameter fitting problem [14,18]. The volume of this parameter space represents the covariance, or uncertainty, of the parameters. Because the FIM is related to the inverse of the covariance matrix [8,10], maximizing the determinant of the FIM results in minimizing the determinant of the covariance matrix, and thus minimizing the volume of the parameter space, or minimizing parameter uncertainty.

The FIM is computed from the sensitivity coefficients of the model, which can be an ODE model as shown in Equation (1). Local sensitivity coefficients are defined as

$$\left. \frac{\partial x_i}{\partial p_j} \right|_{t,p}, \tag{3}$$

where x_i is the i-th state variable in the model, p_j is the j-th model parameter in the vector p, and t is the time at which the partial derivative is evaluated [2,3,32]. The local sensitivity coefficient represents the change in a model state with respect to a change in the value of a parameter, and is a function of time and the parameter vector. At every time point during an experiment, or a model simulation, a sensitivity coefficient can be calculated for any of the state variables. However, the FIM contains the sensitivity coefficients for only those state variables which are measured in experiments for generating data to be used in parameter estimation using least squares fitting, i.e., the sensitivity coefficients for the vector y. The FIM is written as

$$\mathbf{F} = \mathbf{S}^\mathsf{T}\mathbf{S}, \tag{4}$$

where **S** is the sensitivity matrix [32]

$$
\mathbf{S} = \begin{bmatrix}
\frac{\partial y_1}{\partial p_1}(t1) & \frac{\partial y_1}{\partial p_{n_p}}(t1) \\
\vdots & \vdots \\
\frac{\partial y_1}{\partial p_1}(tN) & \frac{\partial y_1}{\partial p_{n_p}}(tN) \\
& \cdots \\
\frac{\partial y_2}{\partial p_1}(t1) & \frac{\partial y_2}{\partial p_{n_p}}(t1) \\
\vdots & \vdots \\
\frac{\partial y_2}{\partial p_1}(tN) & \frac{\partial y_2}{\partial p_{n_p}}(tN)
\end{bmatrix} \in \mathbb{R}^{(2N) x n_p}, \tag{5}
$$

and n_p is the number of parameters in the model. Here, the formula shown for **S** corresponds to an experiment in which two state variables are measured in a time series from time t_1 to t_N; one can extend the formula to the general case of any number of experimentally measured outputs, however, two outputs are realistic for the application investigated in this paper. The local sensitivity coefficients for all of the state variables of the model can be calculated by solving the system of ordinary differential equations

$$
\frac{d}{dt}\frac{\partial x}{\partial p^{\mathsf{T}}} = \frac{\partial f}{\partial x^{\mathsf{T}}}\frac{\partial x}{\partial p^{\mathsf{T}}} + \frac{\partial f}{\partial p^{\mathsf{T}}} \tag{6}
$$

where x is the column vector of state variables, p is the column vector of model parameters, and f is the column vector of functions defining the model, as in Equation (1) [32]. To calculate the D-optimality criterion value for a given experiment design, the model ODEs (Equation (1)) are solved simultaneously with the sensitivity equations (Equation (6)), the FIM is constructed (Equations (4) and (5)), and the determinant can then be computed. To determine an optimal experiment, an optimization problem is solved which searches through different experimental designs, computing the D-optimality criterion in this way for every iteration.

The Morris method [3] can also be utilized for calculating sensitivity coefficients in a covariance matrix. It should be noted that such an approach results in more than local information due to the certain properties of the Morris method. Using the Morris method, global sensitivity coefficients are calculated by repeatedly sampling parameter values from a distribution. Finite difference approximations of the local sensitivities are then computed at each value of the sampled parameter vector, and the finite difference approximations are averaged to obtain each global sensitivity coefficient as shown in Equations (7) and (8).

$$
d_{ijk} = \frac{y_i(t, p_1, \ldots, p_j + \Delta_{jk}, \ldots, p_{n_p}) - y_i(t, p_1, \ldots, p_j, \ldots, p_{n_p})}{\Delta_{jk}} \tag{7}
$$

$$
s_{ij} = \frac{1}{n_d}\sum_{k=1}^{n_d} d_{ijk} \tag{8}
$$

Here, d_{ijk} is a finite difference derivative approximation for the local sensitivity of y_i with respect to parameter p_j at the k-th sampled value of p_j, Δ_{jk} defines the sampled value of p_j depending on the distribution of p_j, and n_d is the number of samples chosen for p_j. This method for calculating sensitivity coefficients takes into account the uncertainty of the parameter values by averaging over samples from the parameter distribution. Therefore, using the Morris method for calculating the sensitivity coefficients in the FIM does not rely on knowledge of an exact value for the parameters. In fact, the exact values of the parameters are by definition unknown—otherwise, one would not need to estimate the parameters. For this problem, the parameter distributions were considered to be normal with a mean at the nominal parameter values and a standard deviation of one-tenth the value of each parameter.

3. Formulation of the Optimal Experimental Design Problem for the IL-6 Model

The values of the parameters p in ODE models are generally unknown and must be estimated using experimentally measured data points [20,30,33]. The problem considered here is to optimize the function for the controlled variable u in order to minimize the uncertainty of the parameters estimated by least squares parameter estimation.

In the IL-6 model, u represents the IL-6 concentration in the local environment of the cells being stimulated by IL-6. The IL-6 concentration at a given time determines the signaling behavior of the cells, which is characterized by the concentrations of each signaling molecule in the pathway at time t. Thus by modulating the input IL-6 concentration, u, as a function of time, the signaling dynamics can be influenced so that measurements of components of the signaling pathway can yield maximal information about the model parameters.

The measured state variables for this problem were chosen to be STAT3N*-STAT3N* and C/EBPβ, two transcription factors of the IL-6 signaling pathway which are directly involved in transcribing DNA to RNA. These measurements were chosen because of the availability of a fluorescent reporter system for measuring the concentrations of these two proteins [2,34]. To design an optimal experiment for minimizing parameter uncertainty, experiments were considered in which the concentrations of these two proteins are measured in a time series every 45 min for 22 h using a fluorescent reporter system. This time sequence for image acquisition was utilized previously to obtain initial estimates for the parameters of the model [2]. However, for the initial parameter estimation, a constant IL-6 concentration at an arbitrary level of 100 ng/mL was utilized as the input function [2]. In the present work, optimal experimental design was applied to optimize the input function over a continuous set of time-dependent input functions while utilizing the same time sequence for data acquisition as was utilized in [2]. This allows for evaluating whether the parameter uncertainty can be decreased by optimizing the input function while holding other experimental control decisions constant.

For this problem, the set of input functions to optimize was chosen to be the piecewise constant input functions with fixed and equal time intervals and IL-6 concentrations bounded between 0 and 7.5 nM. These input functions were chosen because they can be implemented experimentally due to the long intervals during which a concentration is constant, and because they can be parameterized so as to arrive at an optimization problem with a finite number of variables. Specifically, the number of optimization variables is the number of concentration levels allowed in the piecewise constant function. This is illustrated in Figure 1 with a piecewise constant input function for IL-6 that has $r = 4$ concentration levels. As an example, to determine the optimal input function with $r = 4$ concentration levels, the optimal experimental design problem would be solved with four optimization variables representing the four concentration levels of such an input function.

These piecewise constant input functions can be written as

$$
\begin{aligned}
u(t) &= \sum_{k=1}^{r} c_k step(t - (k-1)\Delta t) - \sum_{k=1}^{r} c_k step(t - k\Delta t) \\
&= c_1 step(t) + \sum_{k=2}^{r} [c_k step(t - (k-1)\Delta t) - c_{k-1} step(t - (k-1)\Delta t)] - c_r step(t - r\Delta t) \qquad (9) \\
&= c_1 step(t) + \sum_{k=2}^{r} (c_k - c_{k-1}) step(t - (k-1)\Delta t) - c_r step(t - r\Delta t)
\end{aligned}
$$

where c_k is the k-th concentration level in the vector c, 'step' is the Heaviside step function, r is the number of concentration levels in the input function, and $\Delta t = 22$ h/r is the time interval for each concentration level in the input function.

By changing the vector c to modulate the IL-6 input function, the solution of the model ODEs and local sensitivity ODEs are modulated. This causes the FIM, and thus the D-optimality criterion value, to be a function of c. Therefore, the optimal experimental design problem for minimizing parameter uncertainty in the IL-6 model can be written as

$$\max_{0 \leq c \leq 7.5 \text{ nM}} \ln |\mathbf{F}(c)|, \tag{10}$$

where $|\cdot|$ is the determinant and $\mathbf{F}$ is calculated from Equations (4) and (5) with y_1 from Equation (5) being the concentration of STAT3N*-STAT3N*, y_2 being the concentration of C/EBPβ, and p being the parameters of the IL-6 model. Specifically, in order to calculate the FIM, the IL-6 signaling ODE model is numerically integrated simultaneously with the local sensitivity equations (Equation (6)) for a given input function represented by c, and the FIM is constructed according to Equations (4) and (5). This ODE integration is carried out in every iteration of the optimization problem to evaluate the D-optimality criterion value until the optimization solver determines a solution.

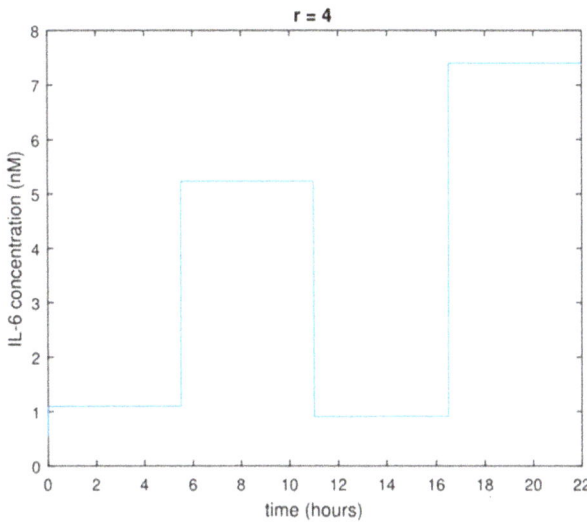

Figure 1. Example of a piecewise constant input function for IL-6 (Interleukin-6) concentration with $r = 4$. The input function shown here can be represented as a vector of the four concentration levels in chronological order: $c = [1.1, 5.2, 0.9, 7.4]$ nM IL-6.

When solving the problem using the Morris method to obtain global sensitivity coefficients, the local sensitivity coefficients in the FIM are replaced by the global sensitivity coefficients calculated by the Morris method. For the Morris method, the local sensitivity ODEs do not need to be solved; rather, the model ODEs are solved repeatedly using the sampled parameter values, and the finite difference approximations of the partial derivatives are computed and averaged.

To solve the optimal experimental design problem for each value of r from 1 to 6, the MATLAB function *'fminsearch'* was utilized [35]. Additionally, to account for the bound constraints on the concentration levels, a stop flag was imposed in the program for running the optimization solver. Multiple initial guesses were utilized for each value of r to avoid local optima (between 5 and 18 initial guesses were used depending on r, in order to cover the range of possible qualitative input function shapes for each value of r).

4. Optimal Experimental Design for the IL-6 Signaling Model

Solving the optimal experimental design problem for minimizing parameter uncertainty in the IL-6 model for values of r from 1 to 6 resulted in an optimal IL-6 input function for each value of r.

The optimal solutions for each *r* are listed in Table 1, which also lists several sub-optimal input functions for comparison. Note that the D-optimality criterion value is greatest for the optimal input function with *r* = 6 (see columns 1–4 in Table 1). The range of D-optimality criterion values for the input functions listed in the table (column 4) is very wide if one considers that the values given are the logarithm of the determinant of the FIM rather than the determinant itself. As expected, the D-optimality criterion values for the optimal input functions increase with *r* (see Table 1, column 4); this is due to the degrees of freedom added when *r* is increased.

Table 1. Optimal and sub-optimal input functions for values of *r* from 1 to 6.

Input Function (nM)	r	Optimal?	D-Optimality Criterion Value	Covariance Norm	Covariance Trace
3.83	1	no	−133.9	161	216
6.59	1	yes	−114.5	43.8	45.2
[0.48, 7.34]	2	yes	−28.9	12.95	12.96
[6.30, 1.60, 7.20]	3	yes	0.30	0.0033	0.0040
[1.09, 5.22, 0.90, 7.39]	4	yes	3.24	6.56×10^{-5}	6.59×10^{-5}
[0.99, 5.64, 0.95, 7.37, 0.97]	5	yes	18.8	1.90×10^{-6}	1.91×10^{-6}
[0.89, 5.94, 1.01, 7.14, 1.07, 5.22]	6	yes	29.7	4.26×10^{-6}	4.83×10^{-6}
[6.97, 0.67, 7.23, 0.88, 7.49, 1.09]	6	no	19.8	0.195	0.204
[7, 6, 5, 4, 3, 2]	6	no	−41.4	42.2	46.8
[1, 4, 6, 6, 4, 1]	6	no	−32.2	1.63	2.03
[3, 4, 5, 6, 7]	5	no	−47.9	55.5	68.8
[6, 6, 5, 3, 3]	5	no	−47.5	1.32	1.44

To test whether higher D-optimality criterion values for an input function corresponded to lower uncertainty, simulations were run for each of the input functions listed in Table 1. For these simulations, data were generated by utilizing the original IL-6 model and adding normally distributed noise to the measurements. Specifically, for a given input function, a simulation was run by integrating the original IL-6 model, adding Gaussian noise with a mean of 0 and a standard deviation of 1 at the measurement time points (every 45 min for 22 h), and fitting the parameters of the simplified IL-6 model to the simulated data. The standard deviation of 1 was chosen to provide a reasonable signal to noise ratio for the model variables. For each input function, 30 simulations were run and parameters were fitted for each simulation. A parameter covariance matrix was then calculated for each of the input functions using the parameter fits from the 30 simulations that were run for each input function. The norm and trace of these covariance matrices were utilized as measures of uncertainty for the parameters. The norm accounts for covariances, while the trace takes into account only variances. Columns 5 and 6 of Table 1 show the norm and trace of the covariance matrix for each input function. As expected, one can observe that the uncertainty is generally lower for optimal input functions than for sub-optimal input functions. However, the relationship between the (*a priori* computed) D-optimality criterion and the (*a posteriori* determined) uncertainty is not monotonic; e.g., the uncertainty for the *r* = 5 optimal input function is slightly lower than that for the *r* = 6 input function while the D-optimality criterion value is greater for the *r* = 6 input function (Table 1). Furthermore, the uncertainty for the *r* = 6 sub-optimal input function shown is larger than that for the optimal *r* = 4 and *r* = 5 input functions even though the D-optimality criterion value is greater for the *r* = 6 sub-optimal input function. There are a number of possible reasons for this lack of monotonicity, such as nonlinearity of the IL-6 model and the fact that optimal experimental design theory is an approximate theory in the case of nonlinear models [8,10].

In order to corroborate that these findings also hold up when global rather than local analysis is used, the Morris method was implemented for computing sensitivity coefficients. This resulted in the optimal input functions listed in Table 2. Figure 2 shows the optimal input functions computed by both the local method and the Morris method for *r* from 1 to 6. It was observed that both methods resulted in optimal input functions with alternating IL-6 concentration levels, reminiscent of pseudo-random binary signals (PRBS signals). While the different methods sometimes result in input profiles that start high vs. low, the general shape of the functions is an oscillation between a low and a high value of the input. This suggests that PRBS-like signals may be favorable for minimizing parameter uncertainty in the IL-6 signaling model.

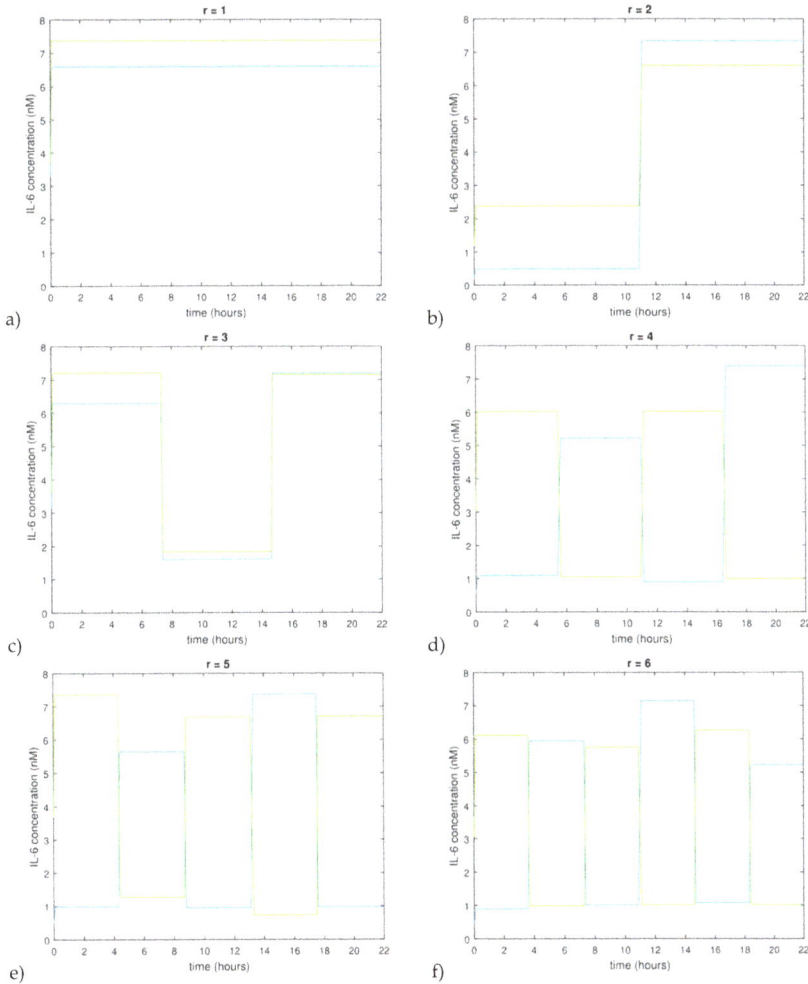

Figure 2. Optimal input functions for values of r from 1 to 6. Blue indicates input functions identified as optimal from the local sensitivity method, green indicates input functions identified as optimal from the Morris method. (**a**) $r = 1$ (**b**) $r = 2$ (**c**) $r = 3$ (**d**) $r = 4$ (**e**) $r = 5$ (**f**) $r = 6$.

Table 2. Optimal input functions for values of r from 1 to 6 via the Morris method

Input Function (nM)	r
7.37	1
[2.38, 6.60]	2
[7.21, 1.83, 7.15]	3
[6.01, 1.05, 6.02, 0.99]	4
[7.36, 1.27, 6.68, 0.73, 6.70]	5
[6.11, 0.99, 5.75, 1.02, 6.26, 1.02]	6

An important take-away message from the simulation results is that optimal experimental design can be very effective at providing an a priori design, yet this does not replace evaluating the quality of a design a posteriori.

5. Discussion

The IL-6 model aims to capture the behavior of molecules involved in both the Jak-STAT pathway and the MAPK pathway [2]. This model is larger and more nonlinear than most models that have previously been the subject of optimal experimental design involving signaling pathways [18,24,25]. The present study shows that optimal experimental design can be utilized for improving the parameter accuracy of moderately complex nonlinear models. Further, the study shows that parameter uncertainty can be substantially decreased by optimizing the input function alone without simultaneously optimizing the experimental measurement time points for this particular system. This observation potentially has fortunate implications for experimental systems in which the input function can be controlled but the experimental measurement time points are dictated by practical considerations and cannot be changed.

The main step in the optimal experimental design was the sensitivity analysis for constructing the FIM. This involves constructing the Jacobian of the model equations so that the sensitivity equations can be integrated to obtain the sensitivity coefficients for the FIM. In principle, this method can be applied to any ODE model. The main step in the analysis is therefore to calculate all of the partial derivatives in the Jacobian, since the Jacobian will be different for every ODE model. For the Morris method, the information matrix is constructed by sampling from the parameter distribution and calculating finite difference derivative approximations for the sensitivities. In both cases—i.e., local and global methods—an ODE solver capable of accurately integrating the ODEs must be used, along with the optimization solver in order to evaluate the objective function for each optimization iteration.

Optimal experimental design using the D-optimality criterion was effective, according to the simulation results, in decreasing the parameter uncertainty in the model. However, one has to caution that this is only one aspect. For example, it has been shown in work by White et al. [36] that increased parameter accuracy does not guarantee an improvement in the accuracy of the model.

The form of the optimal input functions for the IL-6 model raises an interesting point regarding general system identification theory. Each of the determined optimal input functions takes the form of a PRBS-like sequence (see Figure 2), a commonly used input signal for system identification [37], even though this input function shape was not postulated during the formulation of the optimal design problem. Furthermore, computing an experimental design where sensitivities were computed via the Morris method also led to PRBS-like optimal inputs. Without commenting on generality, a PRBS signal seems to be a good choice for inputs of the investigated IL-6 signaling model, a result that may potentially carry over to other signaling pathway models.

6. Conclusions

Optimal experimental design was applied to the problem of minimizing parameter uncertainty in an IL-6 signaling model representing the Jak-STAT and MAPK pathways. The D-optimality criterion, operating on the FIM, was constructed using sensitivity equations, which were solved simultaneously with the equations of the model. Piecewise constant input functions were determined by solving this optimization problem; the piecewise nature of the inputs lays the groundwork for implementing the determined IL-6 concentration profiles on an experimental system. The optimal input functions resulted in decreased parameter uncertainty for the model, as observed from simulations in which parameters were fitted using the optimal input functions for inducing the signaling system. Interestingly, the determined optimal input functions took on the shape of PRBS signals even though this was not in any way postulated by the optimal experimental design problem. This observation was further validated by formulating and solving the problem using a global method in addition to the local method. Future work should corroborate these findings by applying the determined optimal experimental design to an actual experiment.

Acknowledgments: A.S. and J.H. gratefully acknowledge partial financial support from the National Institutes of Health (https://www.nih.gov/, Grant 1R01AI110642).

Author Contributions: A.S. and J.H. conceived and designed the experiments; A.S. performed the experiments and analyzed the data; A.S. and J.H. wrote the paper.

Conflicts of Interest: The authors declare no conflict of interest. The funding sponsors had no role in the design of the study; in the collection, analyses, or interpretation of data; in the writing of the manuscript, or in the decision to publish the results.

Appendix A

The ordinary differential equations comprising the IL-6 model from [2] are shown below, with initial values for the variables and a schematic of the signaling network taken from [2]. Parameter values from the initial fit performed in [2] and values from the average simulated fit using the optimal $r = 5$ input function (from the local method) are shown in Table A1. Units for parameter values correspond to time in seconds and concentrations in nM.

$$dx_1/dt = p_1 u^2 R^2 - p_2 x_1 - p_3 x_1 x_2 + p_4 x_3 + p_5 x_3 - p_{11} x_1 x_5 + p_{12} x_6 - p_{14} x_1 x_7 + p_{15} x_8;$$

$$dx_2/dt = -p_3 x_1 x_2 + p_4 x_3 + 2 p_6 x_4;$$

$$dx_3/dt = p_3 x_1 x_2 - p_4 x_3 - p_5 x_3;$$

$$dx_4/dt = p_5 x_3/2 - p_6 x_4;$$

$$dx_5/dt = p_7 x_4 \, \text{Step}(t - p_8)/(p_9 + x_4) - p_{10} x_5 - p_{11} x_1 x_5 + p_{12} x_6 + p_{13} x_6;$$

$$dx_6/dt = p_{11} x_1 x_5 - p_{12} x_6 - p_{13} x_6;$$

$$dx_7/dt = -p_{14} x_1 x_7 + p_{15} x_8;$$

$$dx_8/dt = p_{14} x_1 x_7 - p_{15} x_8;$$

$$dx_9/dt = p_{16} x_8 x_{10}/(p_{17} + x_{10}) - p_{18} x_9;$$

$$dx_{10}/dt = -p_{16} x_8 x_{10}/(p_{17} + x_{10}) + p_{18} x_9;$$

$$dx_{11}/dt = p_{13} x_6;$$

$$dx_{12}/dt = -2 p_{19} x_9 x_{12}^2;$$

1) $$dx_{13}/dt = p_{19} x_9 x_{12}^2;$$

2)

Name	Component	Initial value (nM)
x_1	(IL6 $-$ gp80 $-$ gp130 $-$ JAK)$_2^*$	0
x_2	STAT3C	1000
x_3	(IL6 $-$ gp80 $-$ gp130 $-$ JAK)$_2^*$ $-$ STAT3C	0
x_4	STAT3N* $-$ STAT3N*	0
x_5	SOCS3	0
x_6	(IL6 $-$ gp80 $-$ gp130 $-$ JAK)$_2^*$ $-$ SOCS3	0
x_7	SHP2	100
x_8	(IL6 $-$ gp80 $-$ gp130 $-$ JAK)$_2^*$ $-$ SHP2 $-$ sum	0
x_9	Erk-PP	0
x_{10}	Erk	16 468
x_{11}	(IL6-gp80-gp130-JAK)$_2$	0
x_{12}	C/EBPβi	40,493
x_{13}	C/EBPβn	0
u	IL-6	3.83 (i.e., 100 ng/ml)
R	Receptor	4

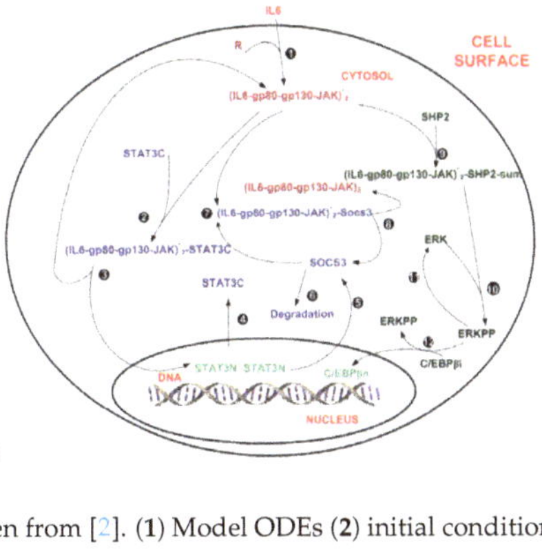

3)

Figure A1. IL-6 model taken from [2]. (**1**) Model ODEs (**2**) initial conditions (**3**) schematic of the IL-6 signaling network.

Table A1. Parameter values from initial parameter fit [2] and from optimal $r = 5$ input function.

Parameter	Initial Fit	$r = 5$ Fit [1]
p_1	2.336×10^{-5}	2.59×10^{-5}
p_2	0.002	0.0128
p_3	0.0138	0.0148
p_4	1.502	1.508
p_5	0.273	0.232
p_6	3.282×10^{-4}	5.653×10^{-4}
p_7	0.023	0.024
p_8	1290	1219
p_9	50.6	52.3
p_{10}	2.067×10^{-4}	5.557×10^{-4}
p_{11}	16.52	16.55
p_{12}	0.04	0.06
p_{13}	0.0023	0.0023
p_{14}	4.059×10^{-4}	4.221×10^{-4}
p_{15}	5.086×10^{-4}	8.717×10^{-4}
p_{16}	16.0	15.9
p_{17}	5.115×10^{3}	5.085×10^{3}
p_{18}	1.198×10^{-5}	1.648×10^{-5}
p_{19}	1.0×10^{-6}	3.0×10^{-5}

[1] Averaged from 30 simulations utilizing the original IL-6 model.

References

1. Singh, A.; Jayaraman, A.; Hahn, J. Modeling Regulatory Mechanisms in IL-6 Signal Transduction in Hepatocytes. *Biotechnol. Bioeng.* **2006**, *95*, 850–862. [CrossRef] [PubMed]
2. Huang, Z.; Chu, Y.; Hahn, J. Model simplification procedure for signal transduction pathway model: An application to IL-6 signaling. *Chem. Eng. Sci.* **2010**, *65*, 1964–1975. [CrossRef]
3. Chu, Y.; Jayaraman, A.; Hahn, J. Parameter sensitivity analysis of IL-6 signalling pathways. *IET Syst. Biol.* **2007**, *1*, 342–352. [CrossRef] [PubMed]
4. Moya, C.; Huang, Z.; Cheng, P.; Jayaraman, A.; Hahn, J. Investigation of IL-6 and IL-10 signalling via mathematical modelling. *IET Syst. Biol.* **2011**, *5*, 15–26. [CrossRef] [PubMed]
5. Himmel, M.E.; Yao, Y.; Orban, P.C.; Steiner, T.S.; Levings, M.K. Regulatory T-cell therapy for inflammatory bowel disease: More questions than answers. *Immunology* **2012**, *136*, 115–122. [CrossRef] [PubMed]
6. Nathan, C. Points of control in inflammation. *Nature* **2002**, *420*, 846–852. [CrossRef] [PubMed]
7. Eastaff-Leung, N.; Mabarrack, N.; Barbour, A.; Cummins, A.; Barry, S. FOXP3+ regulatory T cells, Th17 effector cells, and cytokine environment in inflammatory bowel disease. *J. Clin. Immunol.* **2010**, *30*, 80–89. [CrossRef] [PubMed]
8. Silvey, S.D. *Optimal Design, An Introduction to the Theory for Parameter Estimation*; Chapman and Hall: London, UK, 1980.
9. Fedorov, V.V. *Theory of Optimal Experiments*; Academic Press: New York, NY, USA, 1972.
10. Beck, J.; Arnold, K. *Parameter Estimation in Engineering and Science*; Wiley: Hoboken, NJ, USA, 1977.
11. Atkinson, A.; Donev, A.; Tobias, R. *Optimum Experimental Designs, with SAS*; Oxford University Press: Oxford, UK, 2007.
12. Montgomery, D.C. *Design and Analysis of Experiments*, 5th ed.; Wiley: Hoboken, NJ, USA, 2000.
13. Jones, J.A.; Vernacchio, V.R.; Sinkoe, A.L.; Collins, S.M.; Ibrahim, M.H.A.; Lachance, D.M.; Hahn, J.; Koffas, M.A.G. Experimental and computational optimization of an Escherichia coli co-culture for the efficient production of flavonoids. *Metab. Eng.* **2016**, *35*, 55–63. [CrossRef] [PubMed]
14. Walter, E.; Pronzato, L. Qualitative and Quantitative Experiment Design for Phenomenological Models—A Survey. *Automatica* **1990**, *26*, 195–213. [CrossRef]
15. Wang, H.; Lee, D.; Chen, K.; Chen, J.; Zhang, K.; Silva, A.; Ho, C.; Ho, D. Mechanism-Independent Optimization of Combinatorial Nanodiamond and Unmodified Drug Delivery Using a Phenotypically Driven Platform Technology. *ACS Nano* **2015**, *9*, 3332–3344. [CrossRef] [PubMed]

16. Weiss, A.; Ding, X.; van Beijnum, J.R.; Wong, I.; Wong, T.J.; Berndsen, R.H.; Dormond, O.; Dallinga, M.; Shen, L.; Schlingemann, R.O.; et al. Rapid optimization of drug combinations for the optimal angiostatic treatment of cancer. *Angiogenesis* **2015**, *18*, 233–244. [CrossRef] [PubMed]

17. Bauer, I.; Bock, H.G.; Körkel, S.; Schlöder, J.P. Numerical methods for optimum experimental design in DAE systems. *J. Comput. Appl. Math.* **2000**, *120*, 1–25. [CrossRef]

18. Bandara, S.; Schlöder, J.P.; Eils, R.; Bock, H.G.; Meyer, T. Optimal Experimental Design for Parameter Estimation of a Cell Signaling Model. *PLoS Comput. Biol.* **2009**, *5*, e1000558. [CrossRef] [PubMed]

19. Casey, F.P.; Baird, D.; Feng, Q.; Gutenkunst, R.N.; Waterfall, J.J.; Myers, C.R.; Brown, K.S.; Cerione, R.A.; Sethna, J.P. Optimal experimental design in an epidermal growth factor receptor signalling and down-regulation model. *IET Syst. Biol.* **2007**, *1*, 190–202. [CrossRef] [PubMed]

20. Apgar, J.F.; Witmer, D.K.; White, F.M.; Tidor, B. Sloppy Models, Parameter Uncertainty, and the Role of Experimental Design. *Mol. BioSyst.* **2010**, *6*, 1890–1900. [CrossRef] [PubMed]

21. Balsa-Canto, E.; Alonso, A.A.; Banga, J.R. Computational procedures for optimal experimental design in biological systems. *IET Syst. Biol.* **2008**, *2*, 163–172. [CrossRef] [PubMed]

22. Faller, D.; Klingmüller, U.; Timmer, J. Simulation Methods for Optimal Experimental Design in Systems Biology. *Simulation* **2003**, *79*, 717–725. [CrossRef]

23. Steinmeyer, S.; Howsmon, D.P.; Alaniz, R.C.; Hahn, J.; Jayaraman, A. Empirical modeling of T cell activation predicts interplay of host cytokines and bacterial indole. *Biotechnol. Bioeng.* **2017**. [CrossRef] [PubMed]

24. Chung, M.; Haber, E. Experimental Design for Biological Systems. *SIAM J. Control Optim.* **2012**, *50*, 471–489. [CrossRef]

25. Vanlier, J.; Tiemann, C.A.; Hilbers, P.A.J.; van Riel, N.A.W. A Bayesian approach to targeted experiment design. *Bioinformatics* **2012**, *28*, 1136–1142. [CrossRef] [PubMed]

26. Weber, P.; Kramer, A.; Dingler, C.; Radde, N. Trajectory-oriented Bayesian experiment design versus Fisher A-optimal design: An in depth comparison study. *Bioinformatics* **2012**, *28*, i535–i541. [CrossRef] [PubMed]

27. Silk, D.; Kirk, P.D.W.; Barnes, C.P.; Toni, T.; Stumpf, M.P.H. Model Selection in Systems Biology Depends on Experimental Design. *PLoS Comput. Biol.* **2014**, *10*, e1003650. [CrossRef] [PubMed]

28. Flassig, R.J.; Migal, I.; van der Zalm, E.; Rihko-Struckmann, L.; Sundmacher, K. Rational selection of experimental readout and intervention sites for reducing uncertainties in computational model predictions. *BMC Bioinform.* **2015**, *16*, 13. [CrossRef] [PubMed]

29. Jost, F.; Sager, S.; Thi-Thien Le, T. A Feedback Optimal Control Algorithm with Optimal Measurement Time Points. *Processes* **2017**, *5*, 10. [CrossRef]

30. Marquardt, D.W. An Algorithm for Least-Squares Estimation of Nonlinear Parameters. *J. Soc. Ind. Appl. Math.* **1963**, *11*, 431–441. [CrossRef]

31. Mdluli, T.; Buzzard, G.T.; Rundell, A.E. Efficient Optimization of Stimuli for Model-Based Design of Experiments to Resolve Dynamical Uncertainty. *PLoS Comput. Biol.* **2015**, *11*, e1004488. [CrossRef] [PubMed]

32. Chu, Y.; Hahn, J. Parameter set selection for estimation of nonlinear dynamic systems. *AIChE J.* **2007**, *53*, 2858–2870. [CrossRef]

33. Gutenkunst, R.N.; Waterfall, J.J.; Casey, F.P.; Brown, K.S.; Myers, C.R.; Sethna, J.P. Universally Sloppy Parameter Sensitivities in Systems Biology Models. *PLoS Comput. Biol.* **2007**, *3*, e189. [CrossRef] [PubMed]

34. Huang, Z.; Senocak, F.; Jayaraman, A.; Hahn, J. Integrated Modeling and Experimental Approach for Determining Transcription Factor Profiles from Fluorescent Reporter Data. *BMC Syst. Biol.* **2008**, *2*, 64. [CrossRef] [PubMed]

35. Lagarias, J.C.; Reeds, J.A.; Wright, M.H.; Wright, P.E. Convergence Properties of the Nelder-Mead Simplex Method in Low Dimensions. *SIAM J. Optim.* **1998**, *9*, 112–147. [CrossRef]

36. White, A.; Tolman, M.; Thames, H.D.; Withers, H.R.; Mason, K.A.; Transtrum, M.K. The Limitations of Model-Based Experimental Design and Parameter Estimation in Sloppy Systems. *PLoS Comput. Biol.* **2016**, *12*, e1005227. [CrossRef] [PubMed]

37. Ljung, L. *System Identification: Theory for the User*; Prentice Hall: Upper Saddle River, NJ, USA, 1999.

![processes logo] *processes*

MDPI

Article

Multi-Objective Optimization of Experiments Using Curvature and Fisher Information Matrix

Erica Manesso [1,2,†], **Srinath Sridharan** [3] and **Rudiyanto Gunawan** [1,2,*]

[1]   Institute for Chemical and Bioengineering, ETH Zurich, 8093 Zurich, Switzerland;
      erica.manesso@gmail.com
[2]   Swiss Institute of Bioinformatics, 1015 Lausanne, Switzerland
[3]   Saw Swee Hock School of Public Health, National University of Singapore, Singapore 117549, Singapore;
      srinath@nus.edu.sg
*   Correspondence: rudi.gunawan@chem.ethz.ch; Tel.: +41-44-633-2134
†   Current address: Bayer AG, 65926 Frankfurt am Main, Germany.

Received: 14 September 2017; Accepted: 23 October 2017; Published: 1 November 2017

Abstract: The bottleneck in creating dynamic models of biological networks and processes often lies in estimating unknown kinetic model parameters from experimental data. In this regard, experimental conditions have a strong influence on parameter identifiability and should therefore be optimized to give the maximum information for parameter estimation. Existing model-based design of experiment (MBDOE) methods commonly rely on the Fisher information matrix (FIM) for defining a metric of data informativeness. When the model behavior is highly nonlinear, FIM-based criteria may lead to suboptimal designs, as the FIM only accounts for the linear variation in the model outputs with respect to the parameters. In this work, we developed a multi-objective optimization (MOO) MBDOE, for which the model nonlinearity was taken into consideration through the use of curvature. The proposed MOO MBDOE involved maximizing data informativeness using a FIM-based metric and at the same time minimizing the model curvature. We demonstrated the advantages of the MOO MBDOE over existing FIM-based and other curvature-based MBDOEs in an application to the kinetic modeling of fed-batch fermentation of baker's yeast.

Keywords: design of experiments; multi-objective optimization; Fisher information matrix; curvature; biological processes; mathematical modeling

---

1. Introduction

Dynamic models of biological networks and processes are often created to gain a better understanding of the system behavior. The creation of dynamic biological models requires the values of kinetic parameters, many of which are system-specific and typically not known *a priori*. These parameters are commonly estimated by calibrating model simulations to the available experimental data. Such parameter fitting is known to be challenging, as there often exist multiple parameter combinations that fit the available data equally well; that is, the model parameters are not identifiable [1–5]. While there exist a number of reasons for such lack of parameter identifiability, experimental conditions have a strong influence on this issue and thus should be carefully designed. In addition, biological experiments and data collection are often costly and time-consuming, further motivating the need for well-planned experiments that would give the maximum information given finite resources.

Model-based design of experiments (MBDOEs) offer a means for integrating dynamic modeling with experimental efforts, as illustrated by the iterative procedure in Figure 1. The role of the model here is to capture the knowledge and information about the system up to a given iteration. By using MBDOEs, one could harness this knowledge to guide experiments in the next iteration.

MBDOE techniques have been used extensively for chemical process modeling [6], and more recently, they have been applied to the modeling of cellular processes [7,8]. For the purpose of parameter estimation, experiments are generally designed to improve the precision of the estimated parameters. In this regard, the Fisher information matrix (FIM), whose inverse provides an estimate of the lower bound of parameter variance-covariance by the Cramér-Rao inequality [9], has been commonly used to define the objective function in the optimal experimental design (see [8] and references therein). Since the turn of the century, FIM-based MBDOE methods have had newfound applications in the emerging area of systems biology [10–16]. Besides FIM, Bayesian approaches have also been used for MBDOEs, where given the prior distribution of the model parameters, experiments are designed to minimize the posterior parameter variance [8]. Bayesian MBDOE strategies have been applied to the modeling of biological networks for reducing parametric uncertainty [17–19]. While our work is concerned with MBDOEs for the purpose of parameter estimation, MBDOE strategies have also been developed and applied for discriminating between biological model structures [20–24] and reducing cellular process output uncertainty [19,25,26].

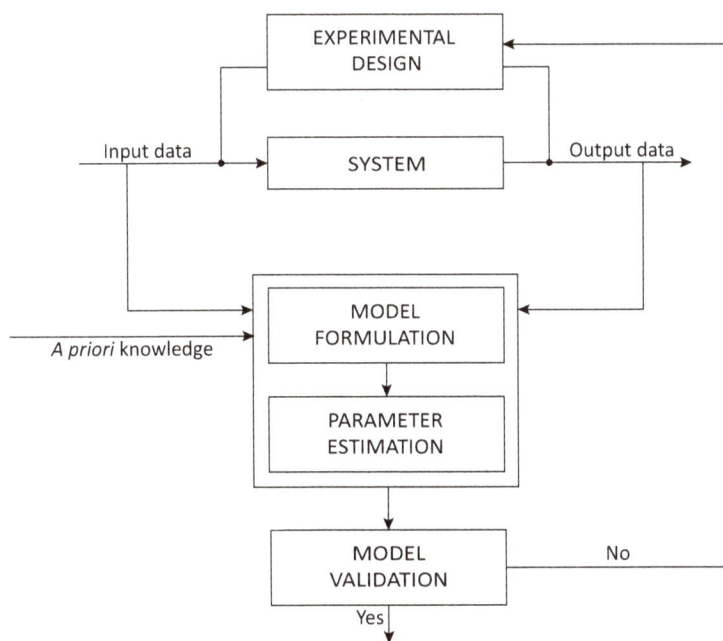

Figure 1. Iterative model identification cycle. The model building process involves the following key steps: experimental design, model structure formulation, parameter estimation, and model validation.

In this work, we focused on FIM-based MBDOEs for parameter estimation. The FIM relies on a linear approximation of the model behavior as a function of the parameters. More precisely, the FIM is computed as a function of the first-order parametric sensitivity coefficients (Jacobian matrix) of model outputs. For systems with a high degree of nonlinearity, the optimal experimental design using the FIM may perform poorly [27]. For this reason, Bates and Watts proposed a MBDOE based on minimizing model curvature by using the second-order parametric sensitivities (Hessian matrix) [28]. Hamilton and Watts further introduced a design criterion, called Q-optimality, based on a quadratic approximation of the volume of the parameter confidence region [29]. More recently, Benabbas et al., proposed two curvature-based MBDOEs [30]. In one design, the authors used a minimization of the root mean square (RMS) of the Hessian matrix, while in another design, they employed a constrained optimization guaranteeing the RMS to be lower than a given level. While the second strategy using a curvature threshold was demonstrated to give more informative experiments, how to set the appropriate RMS threshold value in a particular application was not described.

Recently, Maheshwari et al. described a multi-objective optimization (MOO) formulation for optimizing the design of the experiment using a combination of FIM-based metric and parameter correlation [15]. Because parameter correlations could not account for model nonlinearity, the strategy has the same drawback as FIM-based methods when applied to nonlinear models. In this work, we proposed a MOO MBDOE method using a combination of a FIM criterion and model curvature. We demonstrated the advantages of the proposed MOO MBDOE over FIM-based and other curvature-based methods in an application to the kinetic modeling of the fed-batch fermentation of baker's yeast [30,31].

2. Model-Based Optimal Design of Experiments

We assume that the experimental data $y \in \mathbb{R}^n$ are contaminated by additive random noise, as follows:

$$y = \mu + \epsilon \tag{1}$$

where μ and ϵ denote the mean of the measurement data and the random noise, respectively. When the total number of data points n is greater than the number of parameters p, μ spans a p-dimensional space $\Omega \subset \mathbb{R}^n$, where

$$\Omega = \{\mu : \mu = \mathbf{F}(\mathbf{x}, \mathbf{u}, \boldsymbol{\theta}), \boldsymbol{\theta} \in \Theta \subset \mathbb{R}^p\} \tag{2}$$

Here, $\mathbf{x} \in \mathbb{R}^s$ denotes the state vector, $\boldsymbol{\theta} \in \mathbb{R}^p$ denotes the parameter vector, $\mathbf{u} \in \mathbb{R}^m$ denotes the input and $\mathbf{F}(\mathbf{x}, \mathbf{u}, \boldsymbol{\theta})$ denotes the vector of nonlinear model equations. The subspace Ω is also called the expectation surface or the solution locus. For a dynamic system, the state $\mathbf{x}$ is often described by a set of ordinary differential equations (ODEs):

$$\frac{d\mathbf{x}(\boldsymbol{\theta}, t)}{dt} = \mathbf{g}(\mathbf{x}(\boldsymbol{\theta}, t), \mathbf{u}, \boldsymbol{\theta}), \ \mathbf{x}(\boldsymbol{\theta}, 0) = \mathbf{x}_0 \tag{3}$$

The estimation of model parameters $\boldsymbol{\theta}$ from a given set of data $\mathbf{y}$ is typically formulated as a minimization of the weighted sum of squares of the difference between the model prediction $\mathbf{F}(\mathbf{x}, \mathbf{u}, \boldsymbol{\theta})$ and the measurement data $\mathbf{y}$. For example, the maximum likelihood estimator (MLE) of the model parameters for normally distributed data with known variance $\mathbf{V}$ is given by the minimum of the following objective function:

$$\Phi(\boldsymbol{\theta}) = [\mathbf{y} - \mathbf{F}(\mathbf{x}, \mathbf{u}, \boldsymbol{\theta})]^T \mathbf{V}^{-1} [\mathbf{y} - \mathbf{F}(\mathbf{x}, \mathbf{u}, \boldsymbol{\theta})] \tag{4}$$

When the model is a linear function of the parameters $\mathbf{F}(\mathbf{x}, \mathbf{u}, \boldsymbol{\theta}) = \mathbf{X}\boldsymbol{\theta}$, $\mathbf{X} \in \mathbb{R}^{n \times p}$, then the parameter estimates are given by $\hat{\boldsymbol{\theta}} = (\mathbf{X}^T \mathbf{V}^{-1} \mathbf{X})^{-1} \mathbf{X}^T \mathbf{V}^{-1} \mathbf{y}$. In this case, the MLE is the minimum variance unbiased estimator of θ, where the covariance matrix of the parameter estimates is given by $\mathbf{V}_{\theta} = (\mathbf{X}^T \mathbf{V}^{-1} \mathbf{X})^{-1}$. When the model is nonlinear (with respect to the parameters), the parameter estimates $\hat{\boldsymbol{\theta}} = arg \min \Phi(\boldsymbol{\theta})$ do not necessarily correspond to the minimum variance estimator. According to the Cramér-Rao inequality [9], the inverse of the FIM provides a lower bound for the covariance of the parameter estimates $\hat{\theta}$, that is

$$\mathbf{V}_{\theta} \geq \text{FIM}^{-1} = (\hat{\mathbf{F}}^T \mathbf{V}^{-1} \hat{\mathbf{F}})^{-1} \tag{5}$$

where $\hat{\mathbf{F}} = \dot{\mathbf{F}}(\hat{\boldsymbol{\theta}}, \mathbf{x}) = \frac{\partial \mathbf{F}(\mathbf{x}, \mathbf{u}, \boldsymbol{\theta})}{\partial \boldsymbol{\theta}}|_{\theta = \hat{\theta}}$ is the first-order sensitivity matrix of $\mathbf{F}(\mathbf{x}, \mathbf{u}, \boldsymbol{\theta})$ with respect to the parameters θ.

On the basis of the Cramér-Rao inequality, the FIM has been commonly used as a criterion of data informativeness in MBDOEs. Many methods for MBDOEs, such as those listed in Table 1, are based on finding experimental conditions that optimize a FIM-based information metric. As shown in Equation (5), the FIM relies on a linearization of the model behavior with respect to the parameters. Essentially, the linearization replaces the expectation surface Ω by its tangent plane at $\hat{\theta}$. The performance of the experimental design using a FIM-based criterion would therefore depend on whether (1) the model outputs vary proportionally with the parameter values (planar assumption), and (2) whether this proportionality is constant (uniform coordinate assumption) [32]. When the model is highly nonlinear with respect to the parameters, FIM-based MBDOEs may produce suboptimal designs [33,34]. A recent MOO MBDOE using a combination of a FIM criterion and parameter correlation has been shown to provide an improvement over FIM-based MBDOE methods [15]. However, this method also relies on the first-order parametric sensitivity matrix, and thus it could not account for model nonlinearity.

Table 1. Model-based designs of experiments (MBDOEs) using the Fisher information matrix (FIM).

FIM-Based MBDOE	Criterion
D-optimal	$\max \prod_i \lambda_i$
A-optimal	$\max \sum_i \lambda_i$
E-optimal	$\max \, min(\lambda_i)$
Modified E-optimal	$\max \, \frac{min(\lambda_i)}{max(\lambda_i)}$

Curvature-based design of experiment methods such as the Q-optimality have been introduced to account for model nonlinearity by employing a second-order approximation of the model output. Here, the curvature of the expectation surface Ω is captured using the second-order sensitivities of $F(x, u, \theta)$ based on the Taylor series expansion:

$$F(x, u, \theta) = F(x, u, \hat{\theta}) + \hat{F}(\theta - \hat{\theta}) + \frac{1}{2}(\theta - \hat{\theta})^T \hat{F}(\theta - \hat{\theta}) + O((\theta - \hat{\theta})^3) \qquad (6)$$

where $\hat{F}_{ijk} = \frac{\partial^2 F_i(x,u,\theta)}{\partial \theta_j \partial \theta_k}|_{\theta=\hat{\theta}}$ is the $n \times p \times p$ Hessian matrix. As mentioned in the Introduction, several curvature-based MBDOE methods are available, for example, by minimizing curvature or using a curvature threshold [30]. In this work, we employed a MOO approach based on curvatures for designing optimal experiments. The basic premise of our MBDOE is to select experimental conditions that maximize the informativeness of data and ensure that the model behaves relatively linearly with respect to the parameters. More specifically, our MBDOE uses two objective functions, the first of which involves the maximization of a FIM-based information metric, and the second of which involves the minimization of relative curvature measures [28]. The second objective function ensures that the FIM can provide a reliable measure of data informativeness.

2.1. Multi-Objective Design of Experiments Based on Curvatures

In this section, we derive the relative curvature measures by following the work of Bates and Watts [28]. We consider an arbitrary straight line in the parameter space passing through $\hat{\theta}$:

$$\theta(b) = \hat{\theta} + b\mathbf{h} \qquad (7)$$

where $\mathbf{h} = [h_1, h_2, \ldots, h_p]$ is a non-zero vector. As the scalar parameter b varies, a curve is traced through the expectation surface, also referred to as the lifted line, according to

$$\mu_\mathbf{h}(b) = \mu(\hat{\theta} + b\mathbf{h}) \qquad (8)$$

The tangent line of this curve at $b = 0$ is given by

$$
\begin{aligned}
\boldsymbol{\mu}_{\mathbf{h}} &= \left[\frac{d\boldsymbol{\mu}_{\mathbf{h}}(b)}{db} \right]_{\theta=\hat{\theta}, b=0} \\
&= \left[\sum_{r=1}^{p} \frac{\partial \mathbf{F}(\mathbf{x}, \mathbf{u}, \boldsymbol{\theta})}{\partial \theta_r} \frac{\partial \theta_r(b)}{\partial b} \right]_{\theta=\hat{\theta}, b=0} \\
&= \hat{\mathbf{F}}\mathbf{h}
\end{aligned}
\tag{9}
$$

The set of all such tangent lines, that is, the column space of $\hat{\mathbf{F}}$, describes the tangent (hyper)plane at $\boldsymbol{\mu}(\hat{\boldsymbol{\theta}})$.

Meanwhile, the curvature measures come from a quadratic approximation of $\boldsymbol{\mu}$. In this case, the acceleration of $\boldsymbol{\mu}(b)$ at $b = 0$ can be written as follows:

$$
\ddot{\boldsymbol{\mu}}_{\mathbf{h}} = \mathbf{h}^T \hat{\ddot{\mathbf{F}}} \mathbf{h} = \sum_{i=1}^{p} \sum_{j=1}^{p} \frac{\partial^2 \mathbf{F}(\mathbf{x}, \mathbf{u}, \boldsymbol{\theta})}{\partial \theta_i \partial \theta_j} h_i h_j
\tag{10}
$$

The acceleration vector $\ddot{\boldsymbol{\mu}}_{\mathbf{h}}$ can be subsequently decomposed into two components:

$$
\ddot{\boldsymbol{\mu}}_{\mathbf{h}} = \ddot{\boldsymbol{\mu}}_{\mathbf{h}}^{t} + \ddot{\boldsymbol{\mu}}_{\mathbf{h}}^{n}
\tag{11}
$$

where at $\boldsymbol{\mu}(\hat{\boldsymbol{\theta}})$, $\ddot{\boldsymbol{\mu}}_{\mathbf{h}}^{t}$ is tangential to the tangent plane and $\ddot{\boldsymbol{\mu}}_{\mathbf{h}}^{n}$ is normal to the tangent plane. The tangential acceleration $\ddot{\boldsymbol{\mu}}_{\mathbf{h}}^{t}$ is also called the parameter-effect curvature [28] and provides a measure of nonlinearity along the parameter vector $\mathbf{h}$. The degree of the parameter-effect curvature can change upon reparameterization of the model. Meanwhile, the normal acceleration $\ddot{\boldsymbol{\mu}}_{\mathbf{h}}^{n}$ does not vary with model parameterization, and hence it is called the intrinsic curvature. Finally, the relative curvature measures in the direction of $\mathbf{h}$ are given by [28,32]:

$$
K_{\mathbf{h}}^{t} = \frac{\| \ddot{\boldsymbol{\mu}}_{\mathbf{h}}^{t} \|}{\| \dot{\boldsymbol{\mu}}_{\mathbf{h}}^{t} \|^2}
\tag{12}
$$

$$
K_{\mathbf{h}}^{n} = \frac{\| \ddot{\boldsymbol{\mu}}_{\mathbf{h}}^{n} \|}{\| \dot{\boldsymbol{\mu}}_{\mathbf{h}}^{n} \|^2}
\tag{13}
$$

Below, we describe the decomposition of the Hessian into the tangential and the normal component. We consider the QR-factorization of the Jacobian $\hat{\mathbf{F}}$, that is, $\hat{\mathbf{F}} = \mathbf{QR} = \mathbf{Q} \begin{bmatrix} \tilde{\mathbf{R}} \\ \mathbf{0} \end{bmatrix}$. By rotating the parameter axes $(\boldsymbol{\theta} - \hat{\boldsymbol{\theta}})$ into $\boldsymbol{\varphi} = \tilde{\mathbf{R}}(\boldsymbol{\theta} - \hat{\boldsymbol{\theta}})$, a new Jacobian matrix $\dot{\mathbf{U}} = \frac{d\mathbf{F}(\mathbf{x},\mathbf{u},\boldsymbol{\varphi})}{d\boldsymbol{\varphi}}|_{\boldsymbol{\varphi}=0}$ can be computed as $\dot{\mathbf{U}} = \hat{\mathbf{F}} \tilde{\mathbf{R}}^{-1}$, which comprises the first p column vectors of $\mathbf{Q}$ (i.e., $\mathbf{Q} = \begin{bmatrix} \dot{\mathbf{U}} & \mathbf{N} \end{bmatrix}$). The remaining column vectors of $\mathbf{Q}$ (i.e., $\mathbf{N}$) are orthonormal to the tangent surface at $\boldsymbol{\varphi} = 0$. In the same manner, the Hessian matrix in the rotated axes can be written as $\ddot{\mathbf{U}} = \mathbf{L}^T \hat{\ddot{\mathbf{F}}} \mathbf{L}$, where $\mathbf{L} = \tilde{\mathbf{R}}^{-1}$ and $\ddot{U}_{ijk} = \frac{\partial^2 F_i(\mathbf{x},\mathbf{u},\boldsymbol{\varphi})}{\partial \varphi_j \varphi_k}|_{\boldsymbol{\varphi}=0}$. The decomposition of the Hessian into the tangential and normal components is given by the following equation [28]:

$$
\ddot{\mathbf{A}} = \mathbf{Q}^T \ddot{\mathbf{U}} = \begin{bmatrix} \dot{\mathbf{U}} & \mathbf{N} \end{bmatrix}^T \ddot{\mathbf{U}} = \begin{bmatrix} \ddot{\mathbf{A}}^t & \ddot{\mathbf{A}}^n \end{bmatrix}
\tag{14}
$$

The matrices $\ddot{\mathbf{A}}^t$ and $\ddot{\mathbf{A}}^n$ respectively correspond to the parameter-effect and intrinsic curvature components of the Hessian.

To normalize the relative curvatures in Equations (12) and (13), Bates and Watts [28] used the scaling factor ρ, where $\rho = s\sqrt{p}$ and $s^2 = \frac{(\mathbf{y}-\hat{\boldsymbol{\mu}})^T(\mathbf{y}-\hat{\boldsymbol{\mu}})}{n-p}$. Following the same procedure, we define the normalized relative curvatures as follows:

$$\gamma_{\mathbf{h}}^t = \rho K_{\mathbf{h}}^t \tag{15}$$

$$\gamma_{\mathbf{h}}^n = \rho K_{\mathbf{h}}^n \tag{16}$$

In addition, recasting $\mathbf{h}$ in the rotated axes as $\mathbf{h} = \mathbf{Ld}$, the tangent line $\dot{\boldsymbol{\mu}}_{\mathbf{Ld}}$ will have a unit norm (i.e., $\|\dot{\boldsymbol{\mu}}_{\mathbf{Ld}}\| = 1$) when $\mathbf{d}$ is a unit vector. The computation of $\gamma_{\mathbf{h}}^t$ and $\gamma_{\mathbf{h}}^n$ is thus simplified into

$$\gamma_{\mathbf{Ld}}^t = \rho \|\mathbf{d}^T\ddot{\mathbf{A}}^t\mathbf{d}\|, \forall \mathbf{d} : \|\mathbf{d}\| = 1 \tag{17}$$

$$\gamma_{\mathbf{Ld}}^n = \rho \|\mathbf{d}^T\ddot{\mathbf{A}}^n\mathbf{d}\|, \forall \mathbf{d} : \|\mathbf{d}\| = 1 \tag{18}$$

In the proposed experimental design, the maximum of these curvature measures are used, where

$$\gamma_{max}^t = \max_{\|\mathbf{d}\|=1} \gamma_{\mathbf{Ld}}^t \tag{19}$$

$$\gamma_{max}^n = \max_{\|\mathbf{d}\|=1} \gamma_{\mathbf{Ld}}^n \tag{20}$$

As mentioned above, in formulating the MOO for the design of experiments, two design criteria have been taken into account. The first is that the experiment should be designed to maximize the informativeness of the data for parameter estimation. In this case, we employ an information metric based on the FIM. Meanwhile, the second design criterion in the MOO aims to minimize both the parameter-effect and intrinsic curvatures. The MOO formulation offers certain advantages, for example, that there is no need to prioritize any one of the criteria beforehand. Instead, we generate the Pareto set or Pareto frontier representing the set of solutions for which we cannot improve the value of one objective function without negatively affecting the other(s) [35].

Considering the kinetic ODE model given in Equation (3), our multi-objective formulation using the D-optimal criterion is given by

$$\max_{\mathbf{x}_0, \mathbf{t}_{sp}, \mathbf{u}(t)} \prod_i \lambda_i \tag{21}$$

$$\min_{\mathbf{x}_0, \mathbf{t}_{sp}, \mathbf{u}(t)} \gamma_{max}^t + \gamma_{max}^n$$

subject to

$$\frac{d\mathbf{x}(\hat{\boldsymbol{\theta}}, t)}{dt} = \mathbf{g}(\mathbf{x}(\hat{\boldsymbol{\theta}}, t), \mathbf{u}, \hat{\boldsymbol{\theta}})$$

$$\mathbf{x}(\hat{\boldsymbol{\theta}}, 0) = \mathbf{x}_0 \tag{22}$$

$$\mathbf{x}_0^L \leq \mathbf{x}_0 \leq \mathbf{x}_0^U$$

$$\mathbf{u}_j^L \leq \mathbf{u}_j \leq \mathbf{u}_j^U$$

where λ_i is the ith eigenvalue of the FIM (Equation (5)). The first objective function can be substituted with other FIM-based metrics (see Table 1). The parameter vector $\hat{\boldsymbol{\theta}}$ is either an initial guess of the parameter values or the parameter estimates from the current iteration of an iterative model identification procedure [6]. The decision variables may include the initial condition of the states $\mathbf{x}_0$, the sampling time points of measurements $\mathbf{t}_{sp}$, and the dynamic input $\mathbf{u}(t)$. In the case study below, we considered a control vector parametrization (CVP) of the input $u_i(t)$ as illustrated in Figure 2.

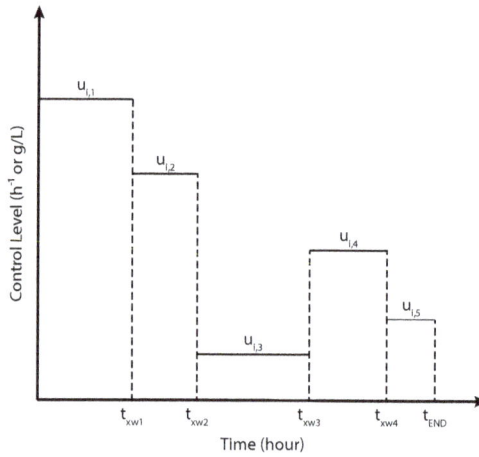

Figure 2. Control vector parametrization of input profiles. In the baker yeast case study, we implemented piecewise constant input profiles with $u_i = [u_{i,1}, u_{i,2}, u_{i,3}, u_{i,4}, u_{i,5}]$ and four switching times: $t_{sw1}, t_{sw2}, t_{sw3}$, and t_{sw4}.

2.2. Numerical Implementation of the Curvature-Based MOO Design

As described in the previous section, the parameter-effect and intrinsic curvatures require the computation of the first- and second-order model sensitivities. For the ODE model in Equation (3), the first-order sensitivities can be calculated according to

$$\hat{\mathbf{F}} = \dot{\mathbf{F}}(\hat{\boldsymbol{\theta}}, \mathbf{x}) = \frac{\partial \mathbf{F}(\mathbf{x}(t, \mathbf{u}, \boldsymbol{\theta}))}{\partial \mathbf{x}} \frac{\partial \mathbf{x}(t, \mathbf{u}, \boldsymbol{\theta})}{\partial \boldsymbol{\theta} / \boldsymbol{\theta}} \bigg|_{\hat{\boldsymbol{\theta}}} \tag{23}$$

The sensitivities in the above equation are normalized with respect to the parameter values. The last term on the right-hand side is the first-order sensitivities of the ODE model, which obey the following differential equation:

$$\frac{d}{dt} \frac{\partial \mathbf{x}}{\partial \boldsymbol{\theta}} = \frac{\partial \mathbf{g}}{\partial \mathbf{x}} \frac{\partial \mathbf{x}}{\partial \boldsymbol{\theta}} + \frac{\partial \mathbf{g}}{\partial \boldsymbol{\theta}}, \quad \frac{\partial \mathbf{x}}{\partial \boldsymbol{\theta}} \bigg|_{t=0} = 0 \tag{24}$$

Here, we have assumed that $\mathbf{x}_0$ is not part of the parameter estimation, but such an assumption can be easily relaxed. In the case study, the sensitivities $\frac{\partial \mathbf{x}}{\partial \boldsymbol{\theta}}$ were computed by solving the ODE in Equation (24) simultaneously to Equation (3), following a procedure known as the direct differential method [36]. Meanwhile, the Hessian matrix was approximated using a finite-difference method, as follows:

$$\hat{\mathbf{F}}_{ijk} = \begin{cases} \frac{F_i(\boldsymbol{\theta} + \Delta\theta_j \mathbf{e}_j) - 2F_i(\boldsymbol{\theta}) + F_i(\boldsymbol{\theta} - \Delta\theta_j \mathbf{e}_j)}{\Delta\theta_j^2 / \theta_j^2}, & \text{for } j = k \\[2ex] \frac{F_i(\boldsymbol{\theta} + \Delta\theta_j \mathbf{e}_j + \Delta\theta_k \mathbf{e}_k) - F_i(\boldsymbol{\theta} + \Delta\theta_j \mathbf{e}_j - \Delta\theta_k \mathbf{e}_k) - F_i(\boldsymbol{\theta} - \Delta\theta_j \mathbf{e}_j + \Delta\theta_k \mathbf{e}_k) + F_i(\boldsymbol{\theta} - \Delta\theta_j \mathbf{e}_j - \Delta\theta_k \mathbf{e}_k)}{(\Delta\theta_j / \theta_j)(\Delta\theta_k / \theta_k)}, & \text{for } j \neq k \end{cases} \tag{25}$$

where $\mathbf{e}_j$ is the jth elementary vector and uses 1% parameter perturbations (i.e., $\Delta\theta_j / \theta_j = 0.01$). The second-order sensitivities above are also normalized with respect to the parameter values.

Meanwhile, the curvature measures γ_{max}^t and γ_{max}^n in Equations (19) and (20) were calculated from the Hessian matrix using the alternating least squares (ALS) method [37], an algorithm created to find the maximum singular value σ_{max} of a three-dimensional matrix. Based on the definitions in Equations (19) and (20), the maximum curvature measures can be determined by computing the maximum singular values of the matrices $\rho\ddot{\mathbf{A}}^t$ and $\rho\ddot{\mathbf{A}}^n$, respectively. More specifically, we implemented the ALS method to solve for

$$\sigma_{max}(\mathbf{B}) = \max_{\|\mathbf{r}\|=\|\mathbf{s}\|=1} \sum_{i=1}^{m} \mathbf{r_i}\,\mathbf{s}^T\mathbf{B_i}\mathbf{s} \tag{26}$$

where $\mathbf{B}$ is either $\rho\ddot{\mathbf{A}}^t$ or $\rho\ddot{\mathbf{A}}^n$. The ALS algorithm started with initial guess values of the vectors $\mathbf{r}$ and $\mathbf{s}$ and used the above equation to solve for one variable while fixing the other in an alternating manner. Zhang and Golub showed that the method linearly converges in a neighbourhood of the optimal solution [37].

In the case study, the MOO problem was solved using the non-dominated sorting genetic algorithm II (NSGAII) in MATLAB, producing a Pareto frontier in the space of the objective functions [38]. We employed a population size of 300 and set the number of generations to 50 times the number of parameters (i.e., 1450). We recasted a maximization of an objective function as the minimization of its negative counterpart. The optimal design was selected from the Pareto frontier by balancing the trade-offs among the objective functions. More specifically, we first normalized the objective functions such that their values on the Pareto frontier ranged between 0 and 1. Finally, we chose among the solutions on the Pareto frontier that which minimized the Euclidean distance of all (normalized) objective functions as the final design.

3. Results

3.1. MBDOEs of Baker Yeast Fermentation Model

We evaluated the performance of the proposed MBDOE in an application to a kinetic model of a fed-batch fermentation of baker's yeast [30,31]. In addition to a D-optimal criterion, we also implemented A-optimal, E-optimal and modified E-optimal criteria (see Table 1) with our MOO MBDOE. We compared the performance of our method to other MBDOEs, including (a) FIM-based MBDOEs, that is, D-optimal, A-optimal, E-optimal and modified E-optimal designs; (b) a D-optimal design with a curvature threshold [30]; (c) a Q-optimal MBDOE [29]; and (d) a MOO MBDOE using parameter correlation [15]. In total, we applied and compared 14 MBDOE methods. For the optimizations in (a), (b) and (c), we employed the enhanced scatter search metaheuristic (eSSm) algorithm [39–41]. For the MOO in (d), we used the optimization algorithm and optimal Pareto point selection, as described in the previous section.

In the fed-batch fermenter model, cellular growth and product formation are captured by the biomass variable x_1, which is assumed to rely on a single substrate variable x_2. The fermenter operates at a constant temperature and the feed is free from product. The model equations are given by

$$\frac{dx_1}{dt} = (r - u_1 - \theta_4)x_1, \; x_1(0) = x_{10}$$
$$\frac{dx_2}{dt} = -\frac{rx_1}{\theta_3} + u_1(u_2 - x_2), \; x_2(0) = 0.1 \tag{27}$$
$$r = \frac{\theta_1 x_2}{\theta_2 + x_2}$$

where the input u_1 is the dilution factor (in the range of 0.05–0.20 h^{-1}) and the input u_2 is the substrate concentration in the feed (in the range of 5–35 g/L). In the model, the biomass growth follows Monod-type kinetics. The parameters θ_1 and θ_2 are the Monod kinetic parameters, θ_3 is the yield coefficient, and θ_4 is the cell death rate constant.

In the MBDOE, the design variables consisted of the initial condition of the biomass $x_1(0)$ in the range between 1 and 10 g/L, 10 measurement sampling times (t_{sp}), and the inputs $u_1(t)$ and $u_2(t)$. The piecewise-constant dynamic inputs were each parametrized using the CVP, as shown in Figure 2. Thus, the MOO was performed with 29 design parameters ($x_1(0)$, 10 t_{sp}'s, 10 $u_{i,j}$'s, and 8 t_{sw}'s). The length of the time interval between two successive measurement sampling points was constrained to be between 1 and 20 h, while that between two input switching times was bounded between 2 and 20 h. The calculations of the Jacobian and Hessian matrices in MBDOEs were made using parameter values $\boldsymbol{\theta}_d = [\theta_1, \theta_2, \theta_3, \theta_4] = [0.5, 0.5, 0.5, 0.5]$ [15,42], which were different from the "true" parameter values used for noisy data generation in the next section. The reason for using a different parameter set in the MBDOE to the true values was to emulate the typical scenario in practice, for which one would start only with an estimate or guess of the model parameters. Figures 3 and 4 show the optimal dynamic inputs and data sampling times resulting from all the MBDOE methods mentioned above (see also the Pareto frontiers in Figures S1 and S2 in the Supplementary Materials). Meanwhile, Table 2 gives the optimal initial biomass concentration $x_1(0)$.

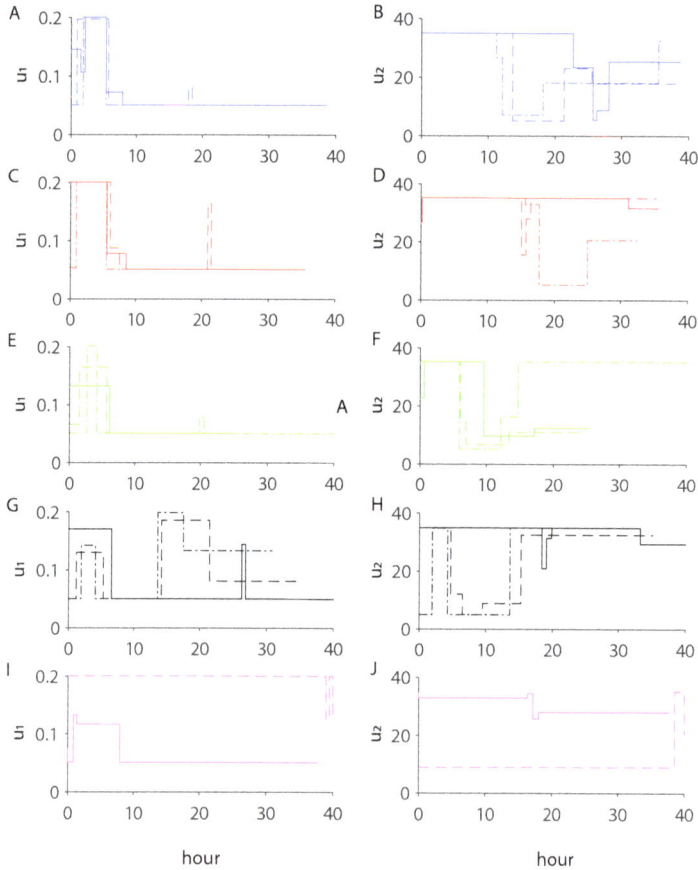

Figure 3. Optimal dilution factor and feed substrate concentration. Optimal dilution factor (u_1 in h^{-1}, left panels) and feed substrate concentration (u_2 in g/L, right panels). (**A,B**) D-optimal (blue). (**C,D**) A-optimal (red). (**E,F**) E-optimal (green). (**G,H**) modified E-optimal (black). In panels (**A–H**), the optimal u_1 and u_2 using Fisher information matrix (FIM)-based criteria are shown by solid line. Those using FIM-based criteria combined with curvatures are shown by dashed line, while those using FIM-based criteria combined with parameter correlation are drawn with dashed-dot line. (**I–J**) Threshold curvature (magenta, solid line), and Q-optimal design (magenta, dashed line).

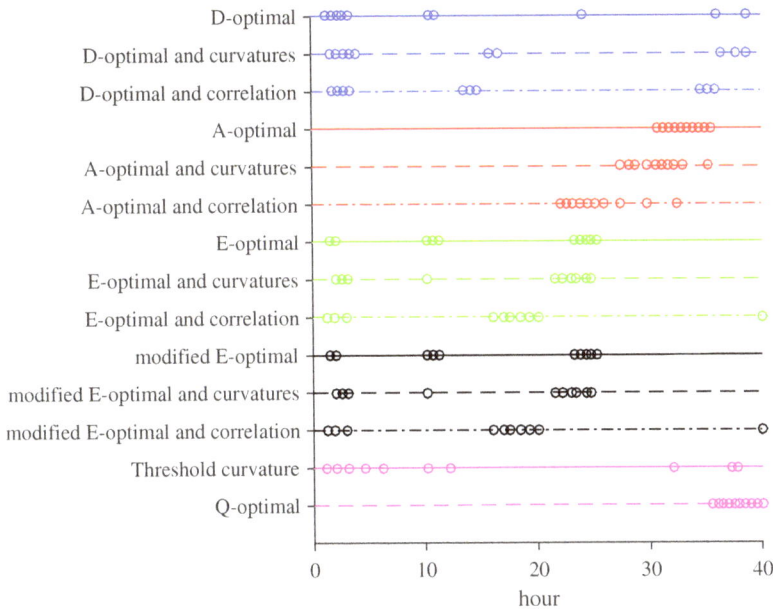

Figure 4. Optimal sampling grid from model-based design of experiments (MBDOEs). Simple Fisher information matrix (FIM)-based criteria shown by continuous line, FIM-based criteria combined with curvatures by dashed line, and FIM-based criteria combined with parameter correlation by dashed-pointed line. Dots indicate the sampling times.

Table 2. Optimal initial condition of biomass $x_1(0)$ (g/L) from model-based design of experiments (MOO: multi-objective optimization).

Design Criterion	$x_1(0)$
D-optimal	10.0
MOO D-optimal and curvatures	10.0
MOO D-optimal and correlation	10.0
A-optimal	10.0
MOO A-optimal and curvatures	9.9
MOO A-optimal and correlation	10.0
E-optimal	10.0
MOO E-optimal and curvatures	10.0
MOO E-optimal and correlation	10.0
Modified E-optimal	10.0
MOO modified E-optimal and curvatures	10.0
MOO modified E-optimal and correlation	10.0
Threshold curvature	8.2
Q-optimal	5.5

3.2. Performance Evaluation

For each of the optimal experimental designs above, we generated in silico datasets by simulating the ODE model using the parameter values $\theta^* = [0.31, 0.18, 0.55, 0.05]$, as reported in previous publications [15,42]. We subsequently added independent and identically distributed (i.i.d.) Gaussian random white noise to the model simulations using a relative variance of 0.04 for both $x_1(t)$ and

$x_2(t)$ [15,42]. For each in silico dataset, we then performed a parameter estimation using the resulting data (y_1 and y_2) by maximum likelihood estimation, that is, by minimizing

$$\Phi(\theta) = \frac{1}{\sigma^2} \sum_{i=1}^{10} [y_1(t_i) - x_1(t_i, \boldsymbol{\theta})]^2 + [y_2(t_i) - x_2(t_i, \boldsymbol{\theta})]^2 \tag{28}$$

We also employed the following constraints for θ in the optimization above:

$$0.05 \le \theta_1, \theta_2, \theta_3 \le 0.98$$

and

$$0.01 \le \theta_4 \le 0.98$$

Finding the globally optimal solution to the parameter estimation in Equation (4) is challenging. Here, we solved the constrained parameter optimization problem using the interior-point algorithm (implemented by the subroutine *fmincon* function in MATLAB) with the true parameter values θ^* as the initial guess. By employing the true values as the initial starting point of the optimization, we expected that the parameter accuracy would mainly be affected by the experimental design and not by the ability of the parameter optimization algorithm to find the globally optimal solution.

We repeated the in silico data generation and parameter estimation as described above 100 times, which resulted in a set of 100 parameter estimates. The performance of each MBDOE was assessed by the average accuracy of the parameter estimates, measured by the average of the normalized mean-square error (nMSE):

$$\overline{\text{nMSE}} = \frac{1}{4} \sum_{i=1}^{4} \text{nMSE}_i \tag{29}$$

where

$$\text{nMSE}_i = \frac{\text{variance}(\hat{\theta}_i) - \text{bias}^2(\hat{\theta}_i)}{(\theta_i^*)^2}, \; i = 1, 2, 3, 4 \tag{30}$$

The variance of $\hat{\theta}_i$ was computed using the set of 100 parameter estimates, while the bias was calculated as the difference between the average of $\hat{\theta}_i$ and θ_i^*. Table 3 gives the average nMSE of the parameter estimates from each MBDOE under consideration.

Table 3. Model-based design of experiment (MBDOE) performance on the fed-batch fermentation of baker's yeast model. The overall parameter accuracy is represented by the average of the normalized mean-square error (nMSE). The reported parameter values and errors are the averages and standard deviations from 100 repeated runs of parameter estimation.

Design Criterion	$\overline{\text{nMSE}}$	$\theta_1 \pm SD_{\theta_1}$	$\theta_2 \pm SD_{\theta_2}$	$\theta_3 \pm SD_{\theta_3}$	$\theta_4 \pm SD_{\theta_4}$
D-optimal	7.06×10^{-3}	0.3107 ± 0.0102	0.1831 ± 0.0276	0.5505 ± 0.0125	0.0502 ± 0.0026
MOO D-optimal and curvatures	4.71×10^{-3}	0.3099 ± 0.0056	0.1825 ± 0.0233	0.5496 ± 0.0099	0.0499 ± 0.0018
MOO D-optimal and correlation	5.36×10^{-3}	0.3117 ± 0.0134	0.1781 ± 0.0151	0.5543 ± 0.0270	0.0508 ± 0.0049
A-optimal	2.35×10^{-1}	0.3294 ± 0.0659	0.2399 ± 0.1387	0.5841 ± 0.1083	0.0558 ± 0.0181
MOO A-optimal and curvatures	1.42	0.3669 ± 0.0947	0.5267 ± 0.2230	0.5548 ± 0.1333	0.0510 ± 0.0244
MOO A-optimal and correlation	4.82	0.0863 ± 0.0499	0.8927 ± 0.2555	0.2879 ± 0.1928	0.0177 ± 0.0263
E-optimal	8.01×10^{-2}	0.3180 ± 0.0420	0.2026 ± 0.0956	0.5473 ± 0.0159	0.0496 ± 0.0026
MOO E-optimal and curvatures	3.33×10^{-3}	0.3083 ± 0.0095	0.1829 ± 0.0164	0.5502 ± 0.0183	0.0500 ± 0.0026
MOO E-optimal and correlation	8.19×10^{-3}	0.3108 ± 0.0164	0.1824 ± 0.0213	0.5552 ± 0.0304	0.0509 ± 0.0055
Modified E-optimal	6.99×10^{-2}	0.3137 ± 0.0165	0.1986 ± 0.0920	0.5498 ± 0.0144	0.0502 ± 0.0033
MOO modified E-optimal and curvatures	3.44×10^{-4}	0.3095 ± 0.0036	0.1789 ± 0.0034	0.5491 ± 0.0073	0.0500 ± 0.0013
MOO modified E-optimal and correlation	2.27×10^{-3}	0.3088 ± 0.0048	0.1820 ± 0.0160	0.5486 ± 0.0047	0.0496 ± 0.0013
Threshold curvature	1.29×10^{-2}	0.3144 ± 0.0307	0.1857 ± 0.0339	0.5500 ± 0.0155	0.0502 ± 0.0032
Q-optimal	1.91×10^{-2}	0.3085 ± 0.0178	0.1757 ± 0.0216	0.5514 ± 0.0236	0.0504 ± 0.0119

4. Discussion

As shown in Figure 3, the MBDOEs prescribed manipulating the input $u_1(t)$ mostly at the beginning of the experiment and the input $u_2(t)$ for the entire duration of the experiment. For the majority of the MBDOEs in this study, the optimal sampling times spread unevenly over the duration of the experiment (see Figure 4). A more detailed comparison between Figures 3 and 4 showed that the optimal sampling points were typically placed before and after a change in the dynamic inputs $u_1(t)$ and $u_2(t)$. The exception to this observation was for the optimal design using the A-optimal criterion, which gave the worst parameter accuracy among the MBDOEs considered.

The consideration of model curvature using the proposed MOO MBDOE generally led to improved parameter accuracy over using only model curvature (i.e., Q-optimal and threshold curvature) or using only FIM-based criteria. The lowest nMSE came from the MOO MBDOE design using the modified E-optimal with model curvature. In comparison to MOO MBDOE using parameter correlation, employing model curvature in the MOO framework gave better experimental designs with lower average nMSEs. Meanwhile, Q-optimality and curvature thresholding strategies provided better nMSEs than the majority of the FIM-based criteria, except the D-optimal design. Finally, the optimal experiments based on the A-optimal criterion, either alone or in MOO MBDOE, performed poorly. The poor performance of the A-optimal design has also been reported in a previous publication [15].

The obvious drawback of curvature-based MBDOEs in comparison to FIM-based strategies is the higher computational cost associated with computing the Hessian matrix. While the number of first-order sensitivities (Jacobian) increases linearly with the number of parameters p, the number of second-order sensitivities scales with p^2. Fortunately, the calculation of the Hessian matrix can be easily parallelized and implemented using multiple computing cores. In practice, one often focuses on only a subset of the model parameters, and therefore the MBDOE is typically done for a handful of parameters.

We note that the MBDOE methods considered in this work consider only parametric uncertainty in the models and assume that uncertainty in model equations, that is, structural uncertainty, is not significant. For certain types of models, such as generalized mass action and S-system models [43,44], model structural uncertainty can be treated as parametric uncertainty, and therefore the MBDOE strategies developed here could be applied. As mentioned in the Introduction, MBDOE methods for discriminating between model structures have been developed, many of which are based on the Bayesian approach. Furthermore, in applications for which there exists intrinsic parametric variability, for example, batch-to-batch variability in cell culture fermentation processes, Bayesian MBDOE methods would be more suitable than FIM-based strategies, as Bayesian methods are able to incorporate the prior (intrinsic) distribution of the parameter values in the design. Nevertheless, as demonstrated in the case study, even when the MOO MBDOEs were performed using model parameters that were quite different from the true values, the resulting optimal designs led to precise and accurate parameter estimates. Meanwhile, biological systems, like other complex systems, have been argued to be sloppy. In the context of our work, sloppy systems lead to mathematical models whose FIMs have eigenvalues that are logarithmically spread evenly over large orders of magnitude [45]. In other words, the system behavior is sensitive to or is controlled by a small number of parameter combinations (along the FIM eigenvectors corresponding to large eigenvalues). At the same time, there exist many parameter combinations that can be varied without affecting the system behavior. Such sloppiness could arise in a system governed by processes that span large and evenly distributed length and/or time scales, such that there exists no clear separation between relevant and irrelevant mechanisms. A recent study demonstrated that in the case of sloppy systems, reducing the model parametric uncertainty by MBDOEs beyond a certain point might not necessarily translate to any improvement in model prediction accuracy [45]. However, it is possible to construct reduced-order models of sloppy systems, whose parameters correspond to the important parameter combinations [46,47]. Parameter estimation and MBDOE strategies can then be applied to these reduced models.

5. Conclusions

Existing MBDOE methods for parameter estimation mostly rely on the FIM to define information criteria. Because the FIM is based on first-order sensitivities with respect to the model parameters, the related MBDOEs may perform poorly for nonlinear models. Here, a new MBDOE using a MOO framework was presented, employing the maximization of a FIM-based information metric and the minimization of model curvatures. The application to a model of the fermentation of baker's yeast demonstrated that accounting model nonlinearity through model curvatures in designing the experiment could lead to improved parameter accuracy over using only a FIM-based criterion. The proposed MOO MBDOE also outperformed other curvature-based designs, including the Q-optimality and curvature thresholding and another MOO MBDOE strategy using parameter correlation. The use of the MOO framework further gives flexibility to accommodate other criteria that may arise in a particular application, in the design of experiments.

Supplementary Materials: The following are available online at http://www.mdpi.com/2227-9717/5/4/63/s1. Figure S1: Pareto frontier of the MOO MBDOE using curvatures and a FIM-based criterion. Figure S2: Pareto frontier of the MOO MBDOE using correlation and a FIM-based criterion.

Acknowledgments: The authors would like to acknowledge funding from ETH Zurich and the Ministry of Education, Singapore.

Author Contributions: R.G. conceived and designed the study; E.M. and S.S. performed the design of the experiments and parameter estimations; E.M., S.S. and R.G. analyzed the data; E.M., S.S. and R.G. wrote the paper.

Conflicts of Interest: The authors declare no conflict of interest. The funding sponsors had no role in the design of the study; in the collection, analyses, or interpretation of data; in the writing of the manuscript; or in the decision to publish the results.

Abbreviations

The following abbreviations are used in this manuscript:

MBDOE	Model-based design of experiments
FIM	Fisher information matrix
MOO	Multi-objective optimization
RMS	Root mean square
ODE	Ordinary differential equation
MLE	Maximum likelihood estimator
CVP	Control vector parametrization
nMSE	Normalized mean-square error

References

1. Srinath, S.; Gunawan, R. Parameter identifiability of power-law biochemical system models. *J. Biotechnol.* **2010**, *149*, 132–140.
2. Jia, G.; Stephanopoulos, G.; Gunawan, R. Ensemble kinetic modeling of metabolic networks from dynamic metabolic profiles. *Metabolites* **2012**, *2*, 891–912.
3. Gábor, A.; Hangos, K.; Banga, J.; Szederkànyi, G. Reaction network realizations of rational biochemical systems and their structural properties. *J. Math. Chem.* **2015**, *53*, 1657–1686.
4. Liu, Y.; Manesso, E.; Gunawan, R. REDEMPTION: Reduced dimension ensemble modeling and parameter estimation. *Bioinformatics* **2015**, *31*, 3387–3389.
5. Villaverde, A.F.; Banga, J.R. Structural properties of dynamic systems biology models: Idenfiability, reachability and initial conditions. *Processes* **2017**, *5*, 29.
6. Franceschini, G.; Macchietto, S. Model-based design of experiments for parameter precision: State of the art. *Chem. Eng. Sci.* **2008**, *63*, 4846–4872.
7. Kreutz, C.; Timmer, J. Systems biology: Experimental design. *FEBS J.* **2009**, *276*, 923–942.
8. Chakrabarty, A.; Buzzard, G.T.; Rundell, A.E. Model-based design of experiments for cellular processes. *WIREs Syst. Biol. Med.* **2013**, *5*, 181–203.

9.   Cover, T.M.; Thomas, J.A. *Elements of Information Theory*, 2nd ed.; John Wiley & Sons: Hoboken, NJ, USA, 2006.
10. Faller, D.; Klingmüller, U.; Timmer, J. Simulation methods for optimal experimental design in systems biology. *Simulation* **2003**, *79*, 717–725.
11. Gadkar, K.G.; Gunawan, R.; Doyle, F.J., III. Iterative approach to model identification of biological networks. *BMC Bioinformatics* **2005**, *6*, 155.
12. Balsa-Canto, E.; Alonso, A.A.; Banga, J.R. Computational procedures for optimal experimental design in biological systems. *IET Syst. Biol.* **2008**, *2*, 163–172.
13. Chung, M.; Haber, E. Experimental design for biological systems. *SIAM J. Control Optim.* **2012**, *50*, 471–489.
14. Transtrum, M.K.; Qiu, P. Optimal experiment selection for parameter estimation in biological differential equation models. *BMC Bioinform.* **2012**, *13*, 181.
15. Maheshwari, V.; Rangaiah, G.P.; Samavedham, L. A multi-objective framework for model based design of experiments to improve parameter precision and minimize parameter correlation. *Ind. Eng. Chem. Res.* **2013**, *52*, 8289–8304.
16. Sinkoe, A.; Hahn, J. Optimal experimental design for parameter estimation of an IL-6 signaling model. *Processes* **2017**, *5*, 49.
17. Vanlier, J.; Tiemann, C.A.; Hilbers, P.A.J.; van Riel, N.A.W. A Bayesian approach to targeted experiment design. *Bioinformatics* **2012**, *28*, 1136–1142.
18. Weber, P.; Kramer, A.; Dingler, C.; Radde, N. Trajectory-oriented Bayesian experiment design versus Fisher A-optimal design: An in-depth comparison study. *Bioinformatics* **2012**, *28*, i535–i541.
19. Liepe, J.; Filippi, S.; Komorowski, M.; Stumpf, M.P.H. Maximizing the information content of experiments in systems biology. *PLoS Comput. Biol.* **2013**, *9*, e1002888.
20. Apgar, J.F.; Toettcher, J.E.; Endy, D.; White, F.M.; Tidor, B. Stimulus design for model selection and validation in cell signaling. *PLoS Comput. Biol.* **2008**, *4*, e30.
21. Daunizeau,J.; Preuschoff, K.; Friston, K.; Stephan, K. Optimizing experimental design for comparing models of brain function. *PLoS Comput. Biol.* **2011**, *7*, e1002280.
22. Flassig, R.J.; Sundmacher, K. Optimal design of stimulus experiments for robust discrimination of biochemical reaction networks. *Bioinformatics* **2012**, *28*, 3089–3096.
23. Busetto, A.G.; Hauser, A.; Krummenacher, G.; Sunnaker, M.; Dimopoulos, S.; Ong, C.S.; Stelling, J.; Buhmann, J.M. Near-optimal experimental design for model selection in systems biology. *Bioinformatics* **2013**, *29*, 2625–2632.
24. Silk, D.; Kirk, P.D.W.; Barnes, C.P.; Toni, T.; Stumpf, M.P.H. Model selection in systems biology depends on experimental design. *PLoS Comput. Biol.* **2014**, *10*, e1003650.
25. Bazil, J.N.; Buzzard, G.T.; Rundell, A.E. A global parallel model based design of experiments method to minimize model output uncertainty. *Bull. Math. Biol.* **2012**, *74*, 688–716.
26. Mdluli, T.; Buzzard, G.T.; Rundell, A.E. Efficient optimization of stimuli for model-based design of experiments to resolve dynamical uncertainty. *PLoS Comput. Biol.* **2015**, *11*, e1004488.
27. Cochran, W.G. Experiments for Nonlinear Functions. *J. Am. Stat. Assoc.* **1973**, *68*, 771–781.
28. Bates, D.M.; Watts, D.G. Relative Curvature Measures of Nonlinearity. *J. R. Stat. Soc. Ser. B* **1980**, *42*, 1–25.
29. Hamilton, D.C.; Watts, D.G. A quadratic design criterion for precise estimation in nonlinear regression models. *Technometrics* **1985**, *27*, 241–250.
30. Benabbas, L.; Asprey, S.P.; Macchietto, S. Curvature-based methods for designing optimally informative experiments in multiresponse nonlinear dynamic situations. *Ind. Eng. Chem. Res.* **2005**, *44*, 7120–7131.
31. Asprey, S.P.; Macchietto, S. Statistical tools for optimal dynamic model building. *Comput. Chem. Eng.* **2000**, *24*, 1261–1267.
32. Seber, G.A.F.; Wild, C.J. *Nonlinear Regression*; John Wiley & Sons: Hoboken, NJ, USA, 2003.
33. Merlé, Y.; Tod, M. Impact of pharmacokinetic-pharmacodynamic model linearization on the accuracy of population information matrix and optimal design. *J. Pharmacokinet. Pharmacodyn.* **2001**, *28*, 363–388.
34. Bogacka, B.; Wright, F. Comparison of two design optimality criteria applied to a nonlinear model. *J. Biopharm. Stat.* **2004**, *14*, 909–930.
35. Rangaiah, G.P. *Multi-Objective Optimization: Techniques and Applications in Chemical Engineering*; World Scientific: Singapore, 2008; Volume 1.
36. Varma, A.; Morbidelli, M.; Wu, H. *Parametric Sensitivity in Chemical Systems*; Cambridge University Press: Cambridge, UK, 1999.

37. Zhang, T.; Golub, G.H. Rank-One Approximation to High Order Tensors. *SIAM J. Matrix Anal. Appl.* **2001**, *23*, 534–550.

38. Deb, K.; Pratap, A.; Agarwal, S.; Meyarivan, T. A fast and elitist multi-objective genetic algorithm: NSGA-II. *IEEE Trans. Evol. Comput.* **2002**, *6*, 182–197.

39. Egea, J.A.; Martí, R.; Banga, J.R. An evolutionary method for complex-process optimization. *Comput. Oper. Res.* **2010**, *37*, 315–324.

40. Egea, J.A.; Rodriguez-Fernandez, M.; Banga, J.R.; Martí, R. Scatter search for chemical and bioprocess optimization. *J. Glob. Optim.* **2007**, *37*, 481–503.

41. Rodriguez-Fernandez, M.; Egea, J.A.; Banga, J.R. Novel metaheuristic for parameter estimation in nonlinear dynamic biological systems. *BMC Bioinform.* **2006**, *7*, 483.

42. Zhang, Y.; Edgar, T.F. PCA combined model-based design of experiments (DOE) criteria for differential and algebraic system parameter estimation. *Ind. Eng. Chem. Res.* **2008**, *47*, 7772–7783.

43. Chou, I.C.; Voit, E. Recent developments in parameter estimation and structure identification of biochemical and genomic systems. *Math. Biosci.* **2009**, *219*, 57–83.

44. Voit, E. Biochemical systems theory: A review. *ISRN Biomath.* **2013**, *2013*, 897658, doi:10.1155/2013/897658.

45. White, A.; Tolman, M.; Thames, H.D.; Withers, H.R.; Mason, K.A.; Transtrum, M.K. The limitations of model-based experimental design and parameter estimation in sloppy systems. *PLoS Comput. Biol.* **2016**, *12*, e1005227.

46. Transturm, M.K.; Qiu, P. Model reduction by manifold boundaries. *Phys. Rev. Lett.* **2014**, *113*, 098701, doi:10.1103/PhysRevLett.113.098701.

47. Transturm, M.K.; Qiu, P. Bridging mechanistic and phenomenological models of complex biological systems. *PLoS Comput. Biol.* **2016**, *12*, e1004915.

processes

Article

Mathematical Modeling of Tuberculosis Granuloma Activation

Steve M. Ruggiero [1,†], **Minu R. Pilvankar** [1,†] **and Ashlee N. Ford Versypt** [1,2,*]

[1] School of Chemical Engineering, Oklahoma State University, Stillwater, OK 74078, USA;
steve.ruggiero@okstate.edu (S.M.R.); minu.pilvankar@okstate.edu (M.R.P.)

[2] Oklahoma Center for Respiratory and Infectious Diseases, Oklahoma State University,
Stillwater, OK 74078, USA

* Correspondence: ashleefv@okstate.edu

† These authors contributed equally to this work.

Received: 23 October 2017; Accepted: 4 December 2017; Published: 11 December 2017

Abstract: Tuberculosis (TB) is one of the most common infectious diseases worldwide. It is estimated that one-third of the world's population is infected with TB. Most have the latent stage of the disease that can later transition to active TB disease. TB is spread by aerosol droplets containing Mycobacterium tuberculosis (Mtb). Mtb bacteria enter through the respiratory system and are attacked by the immune system in the lungs. The bacteria are clustered and contained by macrophages into cellular aggregates called granulomas. These granulomas can hold the bacteria dormant for long periods of time in latent TB. The bacteria can be perturbed from latency to active TB disease in a process called granuloma activation when the granulomas are compromised by other immune response events in a host, such as HIV, cancer, or aging. Dysregulation of matrix metalloproteinase 1 (MMP-1) has been recently implicated in granuloma activation through experimental studies, but the mechanism is not well understood. Animal and human studies currently cannot probe the dynamics of activation, so a computational model is developed to fill this gap. This dynamic mathematical model focuses specifically on the latent to active transition after the initial immune response has successfully formed a granuloma. Bacterial leakage from latent granulomas is successfully simulated in response to the MMP-1 dynamics under several scenarios for granuloma activation.

Keywords: latent tuberculosis; immune system; cytokine signaling network; dynamic systems; collagen remodeling

1. Introduction

Tuberculosis (TB) has killed more people than any other infectious disease and continues to infect more people today than at any other time in history [1]. In 2015, 10.4 million people were infected with Mycobacterium tuberculosis (Mtb), and 1.8 million died from TB disease [2]. An individual inoculated with Mtb may experience a range of outcomes. The Mtb bacteria may be immediately destroyed by the host's immune response, the immune response may isolate bacteria into granulomas where the infection persists in a latent state, or the bacteria may proliferate and manifest as active TB disease if the initial infection is not controlled by the immune response. A majority of people infected with Mtb have a clinically latent infection in which they do not show any symptoms of the infection. These individuals serve as a reservoir for the bacteria, and if their immune response system is compromised in such a way to trigger the penetration of the granulomas by active bacteria and formation of TB cavities, the infection may transition from latent to active TB disease. The major risk factors for activation of TB after an extended latent period include contact with an infectious TB patient, HIV co-infection, initiation of an anti-tumor necrosis factor (TNF) treatment, silicosis, and diabetes [3]. About 5–10% of latent infections undergo granuloma activation and progress to active TB [4]. However, the mechanism for the activation of latent TB is still unclear. Improved understanding of the triggers and dynamics of this transition could be useful for designing new therapies to prevent the activation of latent TB.

TB infection starts when infectious droplets containing Mtb reach the respiratory tract of an individual. After reaching the lung tissue, the Mtb is ingested by the resident alveolar macrophages. The host cellular immune response starts with the secretion of cytokines, such as interleukin 12 (IL-12) and tumor necrosis factor alpha (TNF-α), and chemokines that recruit the immune cells to the site of infection [5] to form a compact cluster of immune cells, known as a granuloma. Latent infection is characterized by granuloma formation and steady state maintenance (Figure 1). A granuloma is mostly comprised of an organized aggregate of blood-derived macrophages that ingest and contain the bacteria, differentiated macrophages, and T cells along with other cells such as neutrophils, multinucleated giant cells, dendritic cells, B cells, natural killer cells, fibroblasts, and cells that secrete extracellular matrix components. The exterior surface of the granuloma is composed largely of collagen fibers (Figure 1). The granuloma acts as a microenvironment that walls off the bacteria from the rest of the body to control the infection [1].

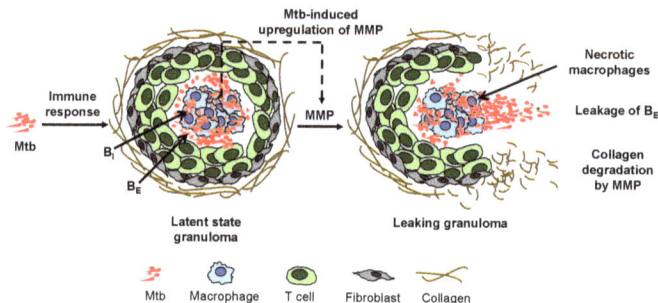

Figure 1. Mycobacterium tuberculosis (Mtb) induces an immune response in the lungs of a host that can lead to formation of a cellular aggregate called a granuloma in which the Mtb can remain dormant in the condition of latent tuberculosis (TB). Direct and indirect upregulation of matrix metalloproteinase (MMP), which is stimulated by the Mtb in infected macrophages in the granuloma and denoted by a dashed arrow, degrades the collagen exterior of the granuloma, triggering leakage of extracellular bacteria (B_E) and formation of a necrotic leaking granuloma called a TB cavity.

Collagen fibers provide tensile strength to the lungs and to granulomas and are highly resistant to enzymatic degradation. Only collagenolytic proteases such as matrix metalloproteinases (MMPs) are able to break the collagen fibers that encapsulate a granuloma. MMPs are a family of proteolytic enzymes that degrade the components of the extracellular matrix and are critical for matrix remodeling [6]. MMPs are typically regulated by the complementary class of inhibitors called tissue inhibitors of metalloproteinases (TIMPs). MMPs have been implicated in the activation of latent tuberculosis infections [7–9]. From the MMP family of proteases, MMP-1 specifically degrades type-1 collagen and drives the remodeling of pulmonary tissue in TB [7]. Experimental data showed that TB activation involved a dysregulation in the balance of MMP-1 and its inhibitor TIMP-1 [7,10]. Direct infection of macrophages induced gene expression and secretion of MMP-1 along with a few other MMPs [7]. Additionally, pro-inflammatory cytokines increased MMP secretion from stromal cells such as epithelial cells, fibroblasts, and astrocytes, while there was no compensatory increase in the production of TIMPs to regulate the MMP levels, thus causing an increase in collagen degradation [7]. Measurements from both human plasma samples [10] and sputum [7] showed that the concentrations of MMP-1 were elevated in patients with active TB compared to latent TB patients or non-infected control subjects, and the levels of the associated inhibitor TIMP-1 were either decreased [7] or changed insignificantly [10] in the active TB cohorts. In [7] MMP-1, degradation of lung collagen in TB was confirmed using a transgenic TB mouse model that overexpressed human MMP-1. As MMP-1 is the dominant MMP in granuloma degradation, we simply refer to MMP-1 as MMP in the rest of this article. A surplus concentration of MMP cleaves the collagen envelope of granulomas and eventually leads to the leakage of bacteria into the airways [8] (Figure 1); from the airways, the bacteria can spread into other regions of the lung and to the rest of the patient's body in an active infection.

The animal models that are most commonly used to study TB do not develop lung pathology exactly the same as in humans [11,12]. Mouse models are useful for studying the infection stage of TB but not the long term latency or reactivation from the latent state because most mouse models do not form human-like granulomas [7]. The rabbit model used in [10] forms necrotic, leaking cavity structures characteristic of active TB after disruption of granulomas. These cavities in rabbits are consistent with the structure observed in humans, providing a valuable animal model for mechanistic insight into cavity formation due to MMP/TIMP imbalance [10]. However, the one month time scale for the formation of active cavities from initial TB infection in the rabbit model is a much faster time scale than the typical cavity formation in human TB infection. Human studies to enhance mechanistic understanding of the untreated latent to active transition cannot be conducted ethically without applying the current standard of care (pharmaceutical interventions), disrupting the cavity formation and progress over time. Additionally, most animal models require that infected animals be sacrificed for invasive lung tissue sample collected to permit observation of the interaction between Mtb and host structures. Thus, conclusions have to be drawn at minimal discrete time points without providing much insight into the dynamic processes. Another surrogate model system is needed to overcome these challenges for understanding the dynamics of MMP dysregulation that can lead to TB cavity formation from latent granulomas and reactivation to TB disease from latent disease.

In lieu of biological experiments, computational models can be used to test possible mechanisms for triggering the switch from latent to active TB infection as well as to study the dynamic process. Mathematical models are useful tools for inexpensively conducting in silico experiments with multiple interacting factors and for testing hypotheses. We developed a mathematical model in this study to probe triggers for inducing the latent to active transition. Several computational and mathematical models have been developed to describe the granuloma formation stages of TB in response to an initial infection [13–16]. Another model was used to explore activation of TB due to a pharmaceutical intervention [17]. No mathematical model has been published addressing the impact of MMP dysregulation or the dynamics of this process on the biological network of cells within a granuloma during TB. The model developed here builds on an existing model of the immune response to Mtb [13] (referred to as the "immune response model" henceforth) by extending this model to explicitly consider dynamic regulation of MMP-1. We also share our open-source Python codes for ease of continued development by other computational researchers and by expanding the horizons of use of the models for further in silico experiments by collaborators and other scientists not necessarily trained in high performance computing.

The immune response model is able to simulate three physiologically-relevant regimes based on parameter values: (i) immediate clearance of Mtb; (ii) a mild initial infection followed by long-term latent TB; and (iii) an initial uncontrolled active infection [13]. Here, a novel model for MMP dynamics, collagen degradation, and bacterial leakage is added to the immune response model. Figure 1 illustrates the mechanisms we aim to capture in the model. The upregulation of MMP by Mtb drives the degradation of the granuloma envelope, which allows Mtb to leak out of the granuloma to the surroundings. Using the model that incorporates the MMP dynamics, we investigated conditions under which the biological system can be perturbed to switch to active infection after a steady latent infection has been established. Section 2 details the equations used to define the mathematical model. Section 3 includes model results under various scenarios as well as an analysis of the model sensitivity to parameter values.

2. Methods

The immune response model considers the local immune response to Mtb in the lungs. The immune response model includes population balances for macrophages, two families of T cells (CD4+ and CD8+), intracellular bacteria (inside infected macrophages), and extracellular bacteria (inside the granuloma but outside the macrophages). The model also includes the signaling network that connects the various cell populations via the cytokines TNF-α, interferon gamma (IFN-γ), IL-4, IL-10, and IL-12 [13]. The equations and parameters of the immune response model are summarized in the Supplementary Materials. The immune response model can generate three regimes representing the infection outcomes of clearance, latency, and active TB. However, the immune response model does not consider the dynamic effects of MMP on the collagen on the surface of the granuloma, which, if breached, can lead to bacterial leakage. In addition, the immune response model cannot predict the transition from latent to active TB after a period of latency. The present work extends the immune response model by considering the effects of intracellular bacteria on reprogramming infected macrophages to increase production of MMP and the subsequent degradation of the collagen envelope of granuloma by MMP. Here, we focus on additions to the original immune response model that represent the local changes in MMP concentration, collagen concentration, and the leakage of extracellular bacteria from the granuloma. Each of these additions is described in turn in the following subsections. Here, the granulomas are considered as well-mixed zones without transport limitations to facilitate adaptation of the ordinary differential equation based immune response model. An alternate partial differential equation model for the immune response with spatial effects including diffusion has been formulated [16]. However, the simulation results for that model were only shown for spatially averaged populations of cells and cytokines. We seek to improve understanding of the process dynamics in the present work and thus follow these previously published models in neglecting spatial effects inside of granulomas.

2.1. MMP Dynamics

The steps that affect the MMP dynamics are illustrated in Figure 2. The activation of resting macrophages, the recruitment of additional macrophages, and the infection of macrophages by Mtb are well-characterized by the immune response model. Macrophages infected with Mtb have been observed to induce gene expression and secretion of MMP; however, a compensatory increase in secretion of TIMP was not observed [7]. The infected macrophages are not the only source of MMP in a granuloma. The infected macrophages interact with the stromal cells like epithelial cells, fibroblasts, and astrocytes, which further secrete MMP and together amplify the MMP upregulation. The pro-inflammatory cytokines especially TNF-α have been found to play a key role in triggering the upregulation of MMPs by stromal cells [18]. It has been found that interaction between macrophages and stromal cell requires TNF-α to increase the MMP secretion by stromal cell networks [18–21].

The mass balance for MMP in terms of concentration for a constant volume system, [MMP], is

$$\frac{d[\text{MMP}]}{dt} = \alpha_{\text{MMP}} M_I \frac{F_\alpha}{F_\alpha + s_{\text{MMP}}} + \beta_{\text{MMP}} M_I - \mu_{\text{MMP}}[\text{MMP}] + sr_{\text{MMP}}, \qquad (1)$$

where the first term represents secretion of MMPs indirectly by the stromal cells that requires both TNF-α, F_α, and infected macrophages, M_I; the second term represents production of MMPs by reprogrammed infected macrophages; the last two terms represent the natural first-order degradation and constant production of MMP; α_{MMP} is the rate constant for indirect production of MMP; s_{MMP} is the constant where the effect of TNF-α on the indirect MMP production has reached half of its saturation level; β_{MMP} is the rate constant for direct production of MMP by M_I; μ_{MMP} is the half life of MMP; and sr_{MMP} is the basal constant recruitment rate of MMP. The last two terms maintain the constant concentration of MMP at equilibrium in the latent state. The functional forms for the two terms representing the upregulation of MMP in the presence of Mtb infection were based on the general mathematical forms defined in [13]: i.e., all terms that require the presence of infected macrophages to upregulate a process are given a linear dependence on M_I, while all terms that are upregulated by a cytokine such as TNF-α are given a Michaelis-Menten type saturation equation dependent on F_α.

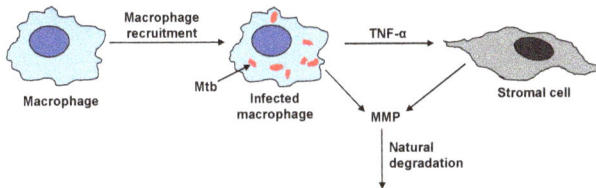

Figure 2. Resting macrophages recruit active macrophages in the immune response. Mtb infects some of the macrophages. The infected macrophages can upregulate MMP secretion directly (denoted by arrow from infected macrophage to MMP) and indirectly via TNF-α signaling to stromal cells (indicated by the arrows connecting the stromal cell to the infected macrophage and MMP). Tissue inhibitors of metalloproteinase (TIMP) is not correspondingly upregulated to inhibit MMP, making the enzyme's degradation rate the primary consumption term for surplus MMP that basal TIMP can not regulate.

2.2. Collagen Dynamics

The collagen dynamics during granuloma formation at the onset of TB infection is beyond the scope of the current work focused on the latent to active transition. Therefore, for simplicity, we consider a constant source term for collagen representative of the source after latency is achieved. The effects of MMP on degrading the collagen are incorporated to study how the granulomas can be compromised after latent TB is established (Figure 3). The cleavage of collagen by MMPs was found to display Michaelis–Menten kinetics [22,23]. We recognize that the granulomas should have the collagen fibers concentrated on the exterior surfaces. In the model proposed here, there is no spatial variation. This could be a realistic approximation if the MMP is uniformly secreted within and adjacent to the granulomas or if the transport occurs faster than the degradation time scale. Slow collagen degradation is considered here, making this well-mixed model reasonable. The mass balance for the change in concentration of collagen, C, in the well-mixed granuloma is

$$\frac{dC}{dt} = sr_C - k_C[\text{MMP}]\frac{C}{C + k_M},\tag{2}$$

where sr_C is a constant recruitment term representing the external build up of the collagen envelope around the granuloma, k_C is the rate constant of collagen degradation, and k_M is the Michaelis constant for collagen degradation catalyzed by MMP.

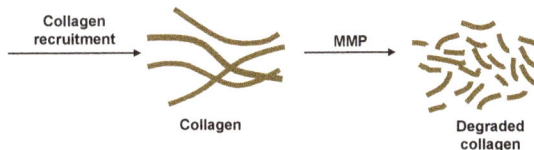

Figure 3. The initial formation of granulomas involves collagen recruitment to form the fibrillar collagen network of the stable granulomas in latent TB. Upregulation of MMP degrades the collagen making the granulomas penetrable by Mtb.

2.3. Bacterial Leakage

The change in the population of the extracellular bacteria inside the granuloma, B_E, has two terms:

$$\frac{dB_E}{dt} = \frac{dB_{E,IR}}{dt} + \frac{dB_{E,L}}{dt}, \tag{3}$$

where $B_{E,IR}$ is the non-leaking extracullular bacteria and $B_{E,L}$ is the leaking extracellular bacteria (Figure 1). The first term $\frac{dB_{E,IR}}{dt}$ is the contribution from the immune response model and is given by (S16) in the Supplementary Materials, which considers all the different mechanisms for the gain and loss in extracellular bacterial count corresponding to the release and uptake of intracellular bacteria by the macrophages, respectively, and the constant turnover number. However, this term does not capture the loss of extracellular bacteria population when the granuloma starts leaking extracellular bacteria into the lung. To account for this case, the second term is added to (3) to track the leakage of bacteria through deteriorated collagen and is represented by

$$\frac{dB_{E,L}}{dt} = -k_L s_L B_E \frac{1 - \dfrac{C}{C_{La}}}{C + s_L}, \tag{4}$$

where k_L is the rate constant for the maximum rate of bacteria exiting the granuloma, s_L is the half saturation constant for the inhibitory effect of collagen on this process, and C_{La} is the expected collagen concentration at latency. When the concentration of collagen is equal to the concentration of collagen in the latent case ($C = C_{La}$), the granuloma is intact with no leakage of bacteria, thus making the leakage term (4) zero. Zero is the maximum value for (4), i.e., the bacterial leakage never has a positive value. This is ensured by the maximum value of C in the model formulation, which is C_{La}. The maximum collagen concentration with respect to changes in MMP concentration is determined by $\frac{\partial C}{\partial [\text{MMP}]} = \frac{\partial C}{\partial t} / \frac{\partial [\text{MMP}]}{\partial t} = 0$. The solution to the maximization problem is $sr_C = k_C [\text{MMP}] \frac{C}{C + k_M}$, which is how sr_C was defined using the values $[\text{MMP}] = [\text{MMP}]_{La}$ and $C = C_{La}$. At the other extreme, when the concentration of collagen goes to zero, there is no longer a barrier around the bacteria, and the rate of leakage is directly dependent on the extracellular bacterial count giving the fastest bacteria leakage rate from the granuloma.

2.4. Biological Feedback

Although not shown explicitly, a feedback loop is formed between the equations introduced here (1)–(4) via the species included in the immune response model (see Supplementary Materials (S1)–(S16)). It is apparent that Equation (2) depends on the value of (MMP), and Equations (3)–(4) depend on C. The contribution to the extracellular bacterial count from the immune response $B_{E,IR}$ depends on multiple species in (S16) including B_E. The extracellular bacteria count B_E in-turn leads directly to changes in species M_R, M_I, M_A, F_α, I_γ, I_{12}, and B_I through dependence on B_E or $B_T = B_E + B_I$ in equations (S1), (S2), (S3), (S10), (S11), (S14), and (S15), respectively. Furthermore, changes in M_I, M_A, I_γ, B_T, and some T cells affect the production of F_α. Changes in M_I and F_α directly lead to changes in the production of MMP given by (1). Other pathways for indirect feedback exist between the cytokines and the bacterial-population-sensitive macrophages.

2.5. Parameter Values

Parameters need to be specified to define the system before performing any simulations. A value of μ_{MMP} is taken from a mathematical model for MMP in fibrosis [24] and is used as the basis for calculating the rest of the parameters for (1). The value of sr_{MMP} is calculated by evaluating (1) with no infection and data from [25]. The value of F_a at the end of a typical latent simulation is used to calculate s_{MMP}. Values of α_{MMP} and β_{MMP} are then calculated using data from [7,25]. Both k_C and k_M are kinetic parameters taken from published experimental data on characterizing the kinetics of MMP. However, the parameters for the breakdown of collagen in literature have two sets of parameters: one for each of the two proteins, α-1 and α-2, that compose collagen I [23]. For this model, a weighted average of the parameters based on the number of proteins in each stand is used and converted into the appropriate units. The source term for collagen I, sr_C, is then calculated from a typical collagen concentration [24]. The value of collagen expected in the latent case, C_{La}, is set based on data from the same fibrosis model used for μ_{MMP} [24]. The rest of the parameters in (4), s_C, k_L, and s_L, are calculated to give the steady state and reasonable limiting behavior. The parameters are defined in Table 1. The parameters listed in the Supplementary Materials Table S1 were validated for the immune response model [13].

Table 1. Parameters in the tuberculosis (TB) granuloma activation model (MMP: matrix metalloproteinase; M_I: infected macrophages; TNF: tumor necrosis factor).

Parameter	Description	Value	Units
α_{MMP}	Rate constant for production of MMP by stromal cells	5.75×10^{-10}	$g \cdot cm^{-3} \cdot M_I^{-1}$
β_{MMP}	Rate constant for production of MMP by M_I	4.44×10^{-10}	$g \cdot cm^{-3} \cdot M_I^{-1}$
s_{MMP}	Half-sat constant of the effect of TNF-α on TNF-α dependent MMP production	2×10^{-1}	$pg \cdot cm^{-3}$
μ_{MMP}	Half life of MMP	4.5	day^{-1}
sr_{MMP}	Constant recruitment rate of MMPs	3.2×10^{-9}	$g \cdot cm^{-3} \cdot day^{-1}$
sr_C	Constant recruitment rate of collagen	1.21×10^{-4}	$g \cdot cm^{-3} \cdot day^{-1}$
k_C	Rate constant of collagen degradation	1.41×10^5	day^{-1}
k_M	Half-sat constant of the effect of collagen on collagen degradation	4.289×10^{-3}	$g \cdot cm^{-3}$
k_L	Rate of bacterial leakage at zero collagen	0.1	$B_E \cdot day^{-1}$
s_L	Half-sat constant on the effect of collagen depletion on bacterial leakage	2×10^{-4}	$g \cdot cm^{-3}$
C_{La}	Concentration of collagen at latency	3.62×10^{-4}	$g \cdot cm^{-3}$

2.6. Numerical Methods and Code Repository

The system of ordinary differential equations in the immune response model (see Supplementary Materials (S1)–(S16)) and the TB granuloma active model defined by (1)–(4) was solved with `odeint` solver from the SciPy Integrate Python module, which uses the classic `lsoda` routine from the FORTAN library `odepack`. The default options were used in the solver. The parameter values in Table 1 were used to generate the results in Section 3, unless otherwise indicated. The initial conditions are given in Table 2. To enable code reuse, we wrote the model in Python and shared the code, parameter files, and documentation in an open-source software repository at http://github.com/ashleefv/tbActivationDynamics [26].

Table 2. Initial conditions for each species in the combined immune response and TB granuloma activation model (TNF: tumor necrosis factor; IFN: interferon; IL: interleukin).

Species	Description	Initial Value	Units
M_R	Resting macrophage count	3.0×10^5	Count
M_I	Infected macrophage count	0	Count
M_A	Activated macrophage count	0	Count
T_0	Th0 cells count	0	Count
T_1	Th1 cells count	0	Count
T_2	Th2 cells count	0	Count
T_{80}	T80 cells count	0	Count
T_8	T8 cells count	0	Count
T_c	TC cells count	0	Count
B_I	Intracellular bacteria	0	Count
B_E	Extracellular bacteria introduced by infection	10	Count
$B_{E,IR}$	Extracellular bacteria generated during immune response to infection	0	Count
$B_{E,L}$	Leaking extracellular bacteria	0	Count
F_α	TNF-α concentration	0	$pg \cdot mL^{-1}$
I_γ	IFN-γ concentration	0	$pg \cdot mL^{-1}$
I_4	IL-4 concentration	0	$pg \cdot mL^{-1}$
I_{10}	IL-10 concentration	0	$pg \cdot mL^{-1}$
I_{12}	IL-12 concentration	0	$pg \cdot mL^{-1}$
C	Collagen concentration	0	$g \cdot g^{-3}$
MMP	MMP concentration	7.11×10^{-10}	$g \cdot g^{-3}$

3. Results and Discussion

3.1. Representative Latent Case

The model proposed in Section 2 is able to simulate both latent and active infections depending on the different parameter values selected. We used substantial leakage of extracellular bacteria as the marker for the transition from latent to active infection. Figure 4 shows model results of a representative latent case using the parameter values listed in Table 1 and Table S1 in the Supplementary Materials. This latent case is also leaking bacteria over time with a very small but nonzero amount of bacteria escaping the granuloma. The leaking case eventually stabilizes and tends to a steady state without further bacterial leakage. It should be noted that the cumulative bacterial leakage observed in Figure 4B after 200 days is on a linear scale compared to the log scale in Figure 4A and is not significant compared to the total bacterial count inside the granuloma. Intracellular bacteria are the bacteria inside the infected macrophages. These infected macrophages secrete TNF-α. The MMP increase tracks with the TNF-α increase, except that the oscillations in MMP are damped compared to those for TNF-α. Around 200 days, the MMP concentration passes a threshold that starts to degrade collagen causing the bacteria leakage term to start growing. The entire system starts stabilizing after that due to the feedback processes, eventually leading to a steady latent state.

3.2. Sensitivity Analysis

In the immune response model, the parameters were probed with a global sensitivity analysis. Here, we conducted a local sensitivity analysis on all of the new parameters for the TB granuloma activation model (Table 1) as well as the parameters for the immune response model (Table S1). The nominal set of parameters were those listed in the tables except for a k_C value of 2.82×10^5 day^{-1}, which is double the latent case value and corresponds to a leaking case. The model output of interest was the total bacterial leakage after the three years (1095 days), which is denoted as B_L. The model output with the nominal set of parameters is B_{Lbase}. All of the parameters were changed one at time by

10 scale factors, s_f, ranging from 0.95 to 1.05 in uniform increments, i.e., increases and decreases by 1%, 2%, . . . , 5%. The normalized local sensitivity index, S_{local}, was calculated using

$$S_{local} = \frac{B_{Lbase} - B_L}{B_{Lbase}\left(1 - s_f\right)},$$ (5)

which is the percent change in bacterial leakage divided by the percent change in the parameter value. Positive numbers suggest increasing the parameter increases the bacterial leakage, and negative numbers suggest decreasing the parameter increases the bacterial leakage. Figure 5 contains the results of the local sensitivity averaged for the range of tested scale factors. All of the parameters were investigated, but only those that were at least as sensitive as k_C are shown in Figure 5. These local sensitivity results are consistent with the global sensitivity results in [13]. k_C is the most sensitive of the parameters introduced here.

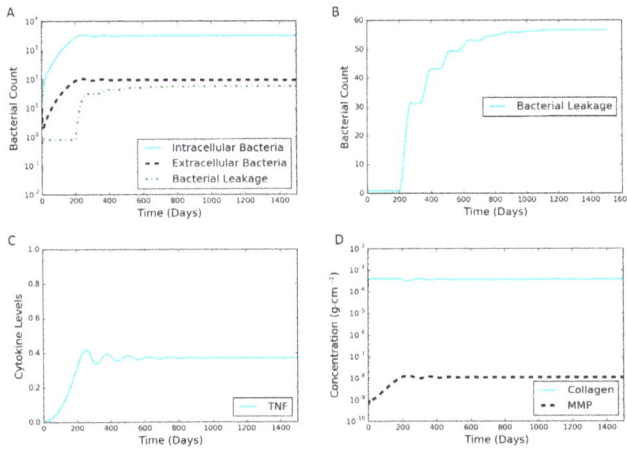

Figure 4. Typical simulation results leading to latent infection. All the concentrations and populations stabilize over time, indicating latency. (**A**) bacterial populations and cumulative bacterial leakage (log scale) vs. time. The intracellular, extracellular, and leaked bacterial concentrations are successfully controlled by the immune system around 200 days; (**B**) cumulative bacterial leakage (linear scale) vs. time; (**C**) dimensionless concentration of cytokine tumor necrosis factor (TNF)-α (linear scale) vs. time; (**D**) concentrations of MMP and collagen vs. time on a log scale.

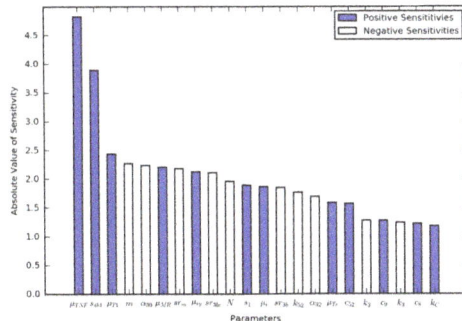

Figure 5. Local sensitivity analysis results for the parameters from the immune response model that were at least as sensitive as the new model parameter k_C.

3.3. Effects of Collagen Degradation Rate Constant k_C

We adjusted the rate constant of collagen degradation, k_C, to determine the effect of the rate of leakage on the other outputs of the model. Not only is k_C the most sensitive of the parameters introduced in Section 2, but it is also the most uncertain of those parameters. There exists a value of k_C where the rate of bacterial leakage is zero, and increasing k_C only affects the other variables of the model through a reduction in collagen allowing more bacteria to leak. Figure 6 contains simulation results at various values of k_C. Increases in k_C have a dampening effect on the oscillations observed in the system (Figure 6A–C) and can lead to a substantial increase in bacterial leakage (Figure 6D).

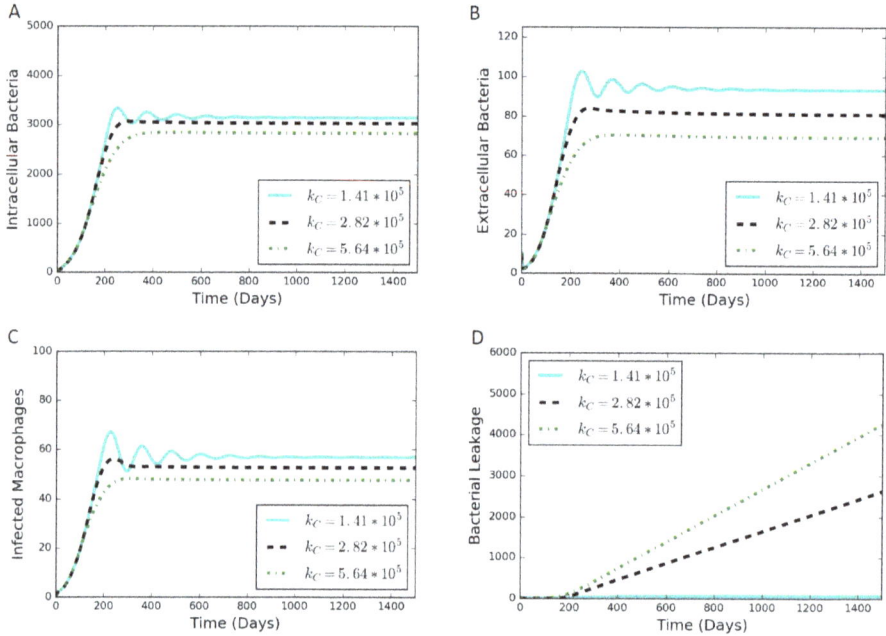

Figure 6. Simulation results from varying values of k_C, where increasing k_C leads towards a leaking state of the granuloma. (**A**) intracellular bacteria count vs. time; (**B**) extracellular bacteria count vs. time; (**C**) infected macrophage count vs. time; (**D**) cumulative bacterial leakage vs. time.

3.4. In Silico Experiment Perturbing the Immune System

An in silico experiment was carried out using the model to examine the effect of perturbing the immune system through the loss of the single immune system components, such as through gene deletion or pharmaceutical interventions. This was conducted by setting the differential equation corresponding to a specific cytokine or cell type to be equal to zero for all time after an initial condition of zero, representing a synthetic suppression of the production of that cell type or cytokine. The rest of the model equations were left unchanged. When starting from the initial conditions and parameters for the typical latent case, one of four results can occur when a specific component of the immune system is not produced during a sustained perturbation of the typical immune response: (1) active infection; (2) formation of a significantly leaking granuloma; (3) a periodic switching between latent and leaking states; and (4) latent infection with little or no leaking. Table 3 summarizes results of the immune system pertubation experiment, and Figure 7 contains results of the immune system pertubation that showcase representative active, leaking, periodic switching, and latent results. The baseline case shown for comparison was the case discussed in Section 3.1 and shown in Figure 4.

Table 3. Summary of in silico immune system pertubation results.

Species Suppressed in the Simulation	Notation	Resulting State
IFN-γ	I_γ	Active
CD4+ Th1 cells	T_1	Active
TNF-α	F_α	Active
CD8+ Tc cells	T_c	Active and highly unstable
CD8+ T8 cells	T_8	Leaking
Activated macrophages	M_A	Leaking
CD4+ Th0 cells	T_0	Periodic switching
CD8+ T80 cells	T_{80}	Periodic switching
IL-12	I_{12}	Periodic Switching
CD4+ Th2 cells	T_2	Latent
IL-4	I_4	Latent
IL-10	I_{10}	Latent

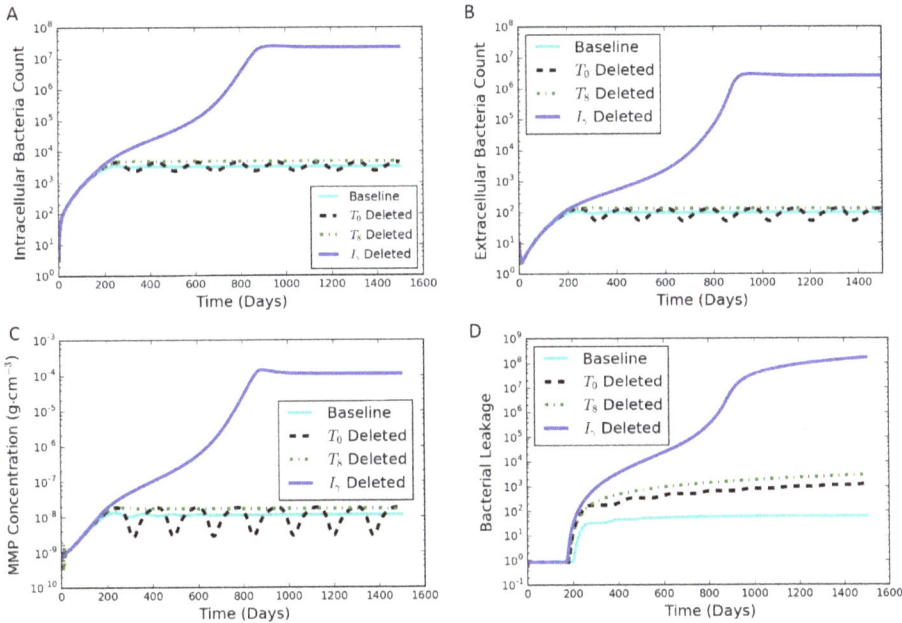

Figure 7. Simulation results of the immune system pertubation experiment where the differential equation corresponding to each of the species shown is set to be zero from day 0, one at a time while the other species follow the otherwise unmodified model equations. Perturbed responses are shown for T_0, T_8, and I_γ (dashed, dotted, and dark solid lines, respectively) The baseline is considered to be a steady latent state (light solid line). (**A**) intracellular bacterial count vs. time; (**B**) extracellular bacterial count vs. time; (**C**) MMP concentration vs. time; (**D**) cumulative bacterial leakage vs. time.

Four species resulted in the active state when they were suppressed: I_γ, T_1, F_α, and T_C. The corresponding intracellular, extracellular and bacterial leakage counts for I_γ, T_1, and F_α increased to the order of 10^8 (Figure 7A,B,D for I_γ), demonstrating an active infection. The results for T_c suppression (not shown) quickly exploded the bacterial and infected macrophages count and never stabilized as in the other three cases.

Eliminating either T_8 (Figure 4) or M_A (not shown because of the similarity to T_8) created a leaking granuloma state characterized by a non-oscillating substantial increase in bacterial leakage (on the order of 10^3). The non-oscillating leakage indicated that the collagen was not able to re-form and control the infection.

We have termed an intermediate state between latent and leaking as "periodic switching" because of sustained oscillations. Eliminating T_0, T_{80}, or I_{12} leads to oscillations in bacterial counts as well as the MMP concentration (Figure 7 for T_0). The constant amplitude of oscillations in the intracellular and extracellular bacterial count and MMP levels suggest that the system was continuously oscillating between two states. For every drop in the MMP levels, the bacterial leakage count stayed steady marking a non-leaking state (flat zone in the stair step pattern of the bacterial leakage curve in Figure 7D). In contrast, as the MMP peaked in every oscillation, it caused a certain constant number of the extracellular bacteria to leak. This led to a corresponding increase in the cumulative bacterial leakage at that point, thus marking a leaking state. The amplitudes of the oscillations for T_0 and T_{80} were similar, while I_{12} had a larger amplitude of oscillations in the range of 20–200 B_E (similar maximum as the other cases but with a lower minimum). The resulting magnitude of bacterial leakage was the same for all three cases, but the duration of the latent periods was longer for I_{12} (not shown).

The simulations for suppressing T_2, I_4, and I_{10} resulted in latent cases (not shown). The results for T_2 suppression were nearly indistinguishable from the baseline case. The bacterial leakage for I_4 was slightly lower than that for the baseline case. The results for I_{10} suppression showed oscillations in B_I and B_E lower than the baseline levels and zero leakage.

3.5. In Silico HIV Co-Infection Experiment via T Cell Depletion

HIV co-infection is known to increase the risk of progression from latent TB to active disease. When a patient is infected with HIV, a major immunological effect is reduction in CD4+ T cells. To model a patient with latent TB becoming co-infected with HIV, an in silico experiment to deplete precursor T_0 cells was performed (Figure 8). The model was run for 1500 days with the baseline case and no changes to the model to establish a latent condition. At 1500 days, parameter α_{1a} was set to zero, and parameters α_2 and sr_{1b} were gradually reduced by an exponentially decaying function for the next 1000 days. All three of these parameters are associated with the number of T_0 cells present within the granuloma (smaller values of the parameters correspond to reduced production and recruitment of T cells). The simulation results show that the system tried to stabilize to until 1500 days (Figure 8) with damping oscillations in T_0 at latency. Shortly after day 1500, there was drop in T_0 cell count for the co-infection. This changed the levels of the cytokines and infected macrophages that affected MMP and collagen concentrations, eventually leading to an increase in bacterial leakage. These results show that a reduction in T_0 cells simulating HIV co-infection after development of latent TB can indeed trigger degradation of collagen and induce a leaking granuloma. The simulation results are consistent with an experimental study in mice co-infected with HIV and Mtb that showed increased mycobacterial burden and dissemination, loss of granuloma structure, and increase progression of TB-disease when the HIV co-infection was present [27]. Another experimental work showed that TB granulomas within HIV-positive expressed more IFN-γ, TNF-α, IL-4, and IL-12 than granulomas from HIV-negative individual [28,29]. Our cytokine results for the HIV co-infection simulation yield gradually increasing levels of IFN-γ, TNF-α, and IL-12 and small decreasing levels of IL-4 (Figure 9).

4. Conclusions

In this study, we extended a model for the immune response to Mtb by adding new equations describing the dynamics of MMP upregulation, collagen degradation, and bacterial leakage. These new equations are able to produce a leaking regime and periodic switching between leaking and non-leaking in addition to the active and latent states that could be modeled with the immune response model. The simulations' results in Section 3 show how MMP–collagen interactions play a significant role in the creation of leaking granulomas, leading to spreading of the infection. Varying the parameter k_c that governs the degradation rate of collagen by MMP had major consequences on the transition from the latent state to the leaking granuloma state. Using this model, we were also able to assess the effects of perturbing the immune response to suppress the responses of various cells and cytokines on the granuloma state in the long-term and the effects of depletion of T_0 cells to simulate HIV co-infection. The model fills a gap in the mathematical modeling of the processes of granuloma activation after latency. The model opens up new directions for computational and experimental studies related to the long-term prognosis of patients with latent TB such as effects of co-infections or vaccines and for further exploration of the dynamics of granuloma activation after latency to improve the understanding of TB disease progression and treatments.

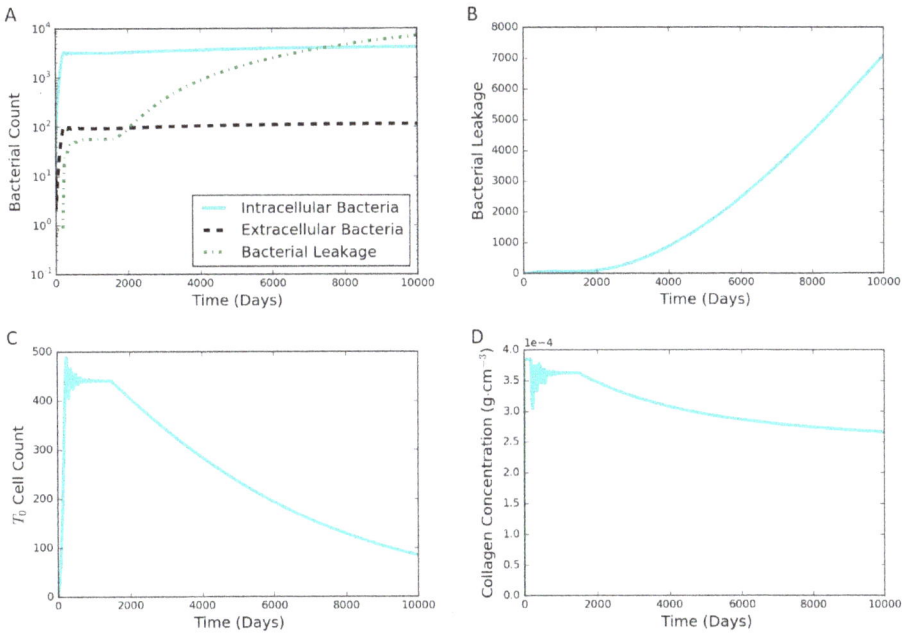

Figure 8. Simulation results for the in silico HIV co-infection experiment via T_0 cell depletion after 1500 days. The depletion results in a leaking state. (**A**) bacterial populations and cumulative bacterial leakage (log scale) vs. time; (**B**) cumulative bacterial leakage (linear scale) vs. time; (**C**) T_0 cell count (linear scale) vs. time; (**D**) concentration of collagen (linear scale) vs. time.

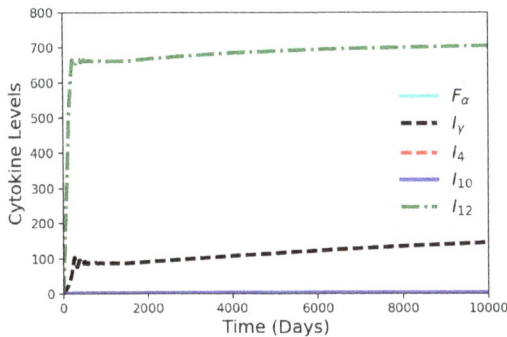

Figure 9. Simulation results for the cytokines from the in silico HIV co-infection experiment via T_0 cell depletion after 1500 days. The depletion results in increasing cytokine levels, except for I_4. The values for F_α, I_{10}, and I_4 are all small.

Supplementary Materials: Equations from the immune response model [13] and the corresponding parameters are available online at http://www.mdpi.com/2227-9717/5/4/79/s1.

Acknowledgments: Research reported in this publication is supported by the National Institute of General Medical Sciences of the National Institutes of Health under Award Number P20GM103648. The content is solely the responsibility of the authors and does not necessarily represent the official views of the National Institutes of Health. Additionally, the authors would like to thank the investigators and visitors of the Oklahoma Center for Respiratory and Infectious Diseases for helpful discussions.

Author Contributions: A.N.F.V. conceived of the study; S.M.R., M.R.P., and A.N.F.V. formulated the model and the numerical solution; S.M.R. and M.R.P. analyzed the results and created visuals; S.M.R., M.R.P., and A.N.F.V. wrote and revised the paper.

Conflicts of Interest: The authors declare no conflict of interest.

References

1. Ramakrishnan, L. Revisiting the role of the granuloma in tuberculosis. *Nat. Rev. Immunol.* **2012**, *12*, 352–366.
2. World Health Organization. *Global Tuberculosis Report 2016*; Technical Report; World Health Organization: Geneva, Switzerland, 2016.
3. Ai, J.W.; Ruan, Q.L.; Liu, Q.H.; Zhang, W.H. Updates on the risk factors for latent tuberculosis reactivation and their managements. *Emerg. Microbes Infect.* **2016**, *5*, e10, doi:10.1038/emi.2016.10.
4. Selwyn, P.A.; Hartel, D.; Lewis, V.A.; Schoenbaum, E.E.; Vermund, S.H.; Klein, R.S.; Walker, A.T.; Friedland, G.H. A prospective study of the risk of tuberculosis among intravenous drug users with human immunodeficiency virus infection. *N. Engl. J. Med.* **1989**, *320*, 545–550.
5. Flynn, J.L.; Chan, J. Immunology of tuberculosis. *Annu. Rev. Immunol.* **2001**, *19*, 93–129.
6. Greenlee, K.J.; Werb, Z.; Kheradmand, F. Matrix metalloproteinases in lung: Multiple, multifarious, and multifaceted. *Physiol. Rev.* **2007**, *87*, 69–98.
7. Elkington, P.; Shiomi, T.; Breen, R.; Nuttall, R.K.; Ugarte-Gil, C.A.; Walker, N.F.; Saraiva, L.; Pedersen, B.; Mauri, F.; Lipman, M.; et al. MMP-1 drives immunopathology in human tuberculosis and transgenic mice. *J. Clin. Investig.* **2011**, *121*, 1827–1833.
8. Salgame, P. MMPs in tuberculosis: Granuloma creators and tissue destroyers. *J. Clin. Investig.* **2011**, *121*, 1686–1688.
9. Sathyamoorthy, T.; Tezera, L.B.; Walker, N.F.; Brilha, S.; Saraiva, L.; Mauri, F.A.; Wilkinson, R.J.; Friedland, J.S.; Elkington, P.T. Membrane type 1 matrix metalloproteinase regulates monocyte migration and collagen destruction in tuberculosis. *J. Immunol.* **2015**, *195*, 882–891.
10. Kubler, A.; Luna, B.; Larsson, C.; Ammerman, N.C.; Andrade, B.B.; Orandle, M.; Bock, K.; Xu, Z.; Bagci, U.; Mollura, D.; et al. Mycobacterium tuberculosis dysregulates MMP/TIMP balance to drive rapid cavitation and unrestrained bacterial proliferation. *J. Pathol.* **2015**, *235*, 431–444.
11. North, R.J.; Jung, Y.J. Immunity to tuberculosis. *Annu. Rev. Immunol.* **2004**, *22*, 599–623.
12. Young, D. Animal models of tuberculosis. *Eur. J. Immunol.* **2009**, *39*, 2011–2014.
13. Sud, D.; Bigbee, C.; Flynn, J.L.; Kirschner, D.E. Contribution of CD8+ T cells to control of *Mycobacterium tuberculosis* infection. *J. Immunol.* **2006**, *176*, 4296–4314.
14. Fallahi-Sichani, M.; El-Kebir, M.; Marino, S.; Kirschner, D.; Linderman, J. Multiscale computational modeling reveals a critical role for TNF-alpha receptor 1 dynamics in tuberculosis granuloma formation. *J. Immunol.* **2011**, *186*, 3472–3483.
15. Cilfone, N.; Perry, C.; Kirschner, D.; Linderman, J. Multi-scale modeling predicts a balance of tumor necrosis factor-alpha and interleukin-10 controls the granuloma environment during *Mycobacterium tuberculosis* infection. *PLoS ONE* **2013**, *8*, e68680, doi:10.1371/journal.pone.0068680.
16. Hao, W.; Schlesinger, L.S.; Friedman, A. Modeling granulomas in response to infection in the Lung. *PLoS ONE* **2016**, *11*, e0148738, doi:10.1371/journal.pone.0148738.
17. Marino, S.; Sud, D.; Plessner, H.; Lin, P.L.; Chan, J.; Flynn, J.L.; Kirschner, D.E. Differences in reactivation of tuberculosis induced from anti-TNF treatments are based on bioavailability in granulomatous tissue. *PLoS Comput. Biol.* **2007**, *3*, 1904–1924.
18. O'Kane, C.M.; Elkington, P.T.; Friedland, J.S. Monocyte-dependent oncostatin M and TNF-α synergize to stimulate unopposed matrix metalloproteinase-1/3 secretion from human lung fibroblasts in tuberculosis. *Eur. J. Immunol.* **2008**, *38*, 1321–1330.

19. Elkington, P.T.; Green, J.A.; Emerson, J.E.; Lopez-Pascua, L.D.; Boyle, J.J.; O'Kane, C.M.; Friedland, J.S. Synergistic up-regulation of epithelial cell matrix metalloproteinase-9 secretion in tuberculosis. *Am. J. Respir. Cell Mol. Biol.* **2007**, *37*, 431–437.

20. Elkington, P.T.; Emerson, J.E.; Lopez-Pascua, L.D.; O'Kane, C.M.; Horncastle, D.E.; Boyle, J.J.; Friedland, J.S. Mycobacterium tuberculosis up-regulates matrix metalloproteinase-1 secretion from human airway epithelial cells via a p38 MAPK switch. *J. Immunol.* **2005**, *175*, 5333–5340.

21. O'Kane, C.M.; Elkington, P.T.; Jones, M.D.; Caviedes, L.; Tovar, M.; Gilman, R.H.; Stamp, G.; Friedland, J.S. STAT3, p38 MAPK, and NF-κB drive unopposed monocyte-dependent fibroblast MMP-1 secretion in tuberculosis. *Am. J. Respir. Cell Mol. Biol.* **2010**, *43*, 465–474.

22. Turto, H.; Lindy, S.; Uitto, V.J.; Wegelius, O.; Uitto, J. Human leukocyte collagenase: Characterization of enzyme kinetics by a new method. *Anal. Biochem.* **1977**, *83*, 557–569.

23. Fasciglione, G.; Gioia, M.; Tsukada, H.; Liang, J.; Iundusi, R.; Tarantino, U.; Coletta, M.; Pourmotabbed, T.; Marini, S. The collagenolytic action of MMP-1 is regulated by the interaction between the catalytic domain and the hinge region. *J. Biol. Inorg. Chem.* **2012**, *17*, 663–672.

24. Hao, W.; Rovin, B.H.; Friedman, A. Mathematical model of renal interstitial fibrosis. *Proc. Natl. Acad. Sci. USA* **2014**, *111*, 14193–14198.

25. Brace, P.T.; Tezera, L.B.; Bielecka, M.K.; Mellows, T.; Garay, D.; Tian, S.; Rand, L.; Green, J.; Jogai, S.; Steele, A.J.; et al. Mycobacterium tuberculosis subverts negative regulatory pathways in human macrophages to drive immunopathology. *PLoS Pathog.* **2017**, *13*, 1–25.

26. Ruggiero, S.M.; Ford Versypt, A.N. tbActivationDynamics, 2017, doi:10.5281/zenodo.1034561. Available online: http://github.com/ashleefv/tbActivationDynamics (accessed on 21 October 2017).

27. Nusbaum, R.J.; Calderon, V.E.; Huante, M.B.; Sutjita, P.; Vijayakumar, S.; Lancaster, K.L.; Hunter, R.L.; Actor, J.K.; Cirillo, J.D.; Aronson, J.; et al. Pulmonary tuberculosis in humanized mice infected with HIV-1. *Sci. Rep.* **2016**, *6*, 21522, doi:10.1038/srep21522.

28. Bourgarit, A.; Carcelain, G.; Martinez, V.; Lascoux, C.; Delcey, V.; Gicquel, B.; Vicaut, E.; Lagrange, P.H.; Sereni, D.; Autran, B. Explosion of tuberculin-specific Th1-responses induces immune restoration syndrome in tuberculosis and HIV co-infected patients. *AIDS* **2006**, *20*, F1–F7.

29. Dierich, C.R.; Flynn, J.L. HIV-1/Mycobacterium tuberculosis coinfection immunology: How does HIV-1 exacerbate tuberculosis? *Infect. Immun.* **2011**, *79*, 1407–1417.

processes

MDPI

Article

Mathematical Modeling and Parameter Estimation of Intracellular Signaling Pathway: Application to LPS-induced NFκB Activation and TNFα Production in Macrophages

Dongheon Lee [1,2], Yufang Ding [3], Arul Jayaraman[1,3] and Joseph S. Kwon [1,2,]*

[1] Artie McFerrin Department of Chemical Engineering, Texas A&M University,
    College Station, TX 77843, USA; dl9@tamu.edu (D.L.); arulj@tamu.edu (A.J.)
[2] Texas A&M Energy Institute, Texas A&M University, College Station, TX 77843, USA
[3] Department of Biomedical Engineering, Texas A&M University,
    College Station, TX 77843, USA; flamytulip@tamu.edu
*   Correspondence: kwonx075@tamu.edu; Tel.: +1-979-962-5930

Received: 28 January 2018; Accepted: 21 February 2018; Published: 25 February 2018

Abstract: Due to the intrinsic stochasticity, the signaling dynamics in a clonal population of cells exhibit cell-to-cell variability at the single-cell level, which is distinct from the population-average dynamics. Frequently, flow cytometry is widely used to acquire the single-cell level measurements by blocking cytokine secretion with reagents such as Golgiplug™. However, Golgiplug™ can alter the signaling dynamics, causing measurements to be misleading. Hence, we developed a mathematical model to infer the average single-cell dynamics based on the flow cytometry measurements in the presence of Golgiplug™ with lipopolysaccharide (LPS)-induced NFκB signaling as an example. First, a mathematical model was developed based on the prior knowledge. Then, average single-cell dynamics of two key molecules (TNFα and IκBα) in the NFκB signaling pathway were measured through flow cytometry in the presence of Golgiplug™ to validate the model and maximize its prediction accuracy. Specifically, a parameter selection and estimation scheme selected key model parameters and estimated their values. Unsatisfactory results from the parameter estimation guided subsequent experiments and appropriate model improvements, and the refined model was calibrated again through the parameter estimation. The inferred model was able to make predictions that were consistent with the experimental measurements, which will be used to construct a semi-stochastic model in the future.

Keywords: systems biology; parameter estimation; NFκB signaling pathway; lipopolysaccharide; flow cytometry; sensitivity analysis

1. Introduction

To integrate of multiple signaling pathways, their canonical transcription factors and downstream effector genes is required for cells to respond to various signals they encounter in their micro-environment. Therefore, understanding how information is sensed and processed by cells and the signaling pathways that are engaged by different stimuli can help elucidate cellular behaviors and responses. Typically, cellular signal dynamics and the response to stimuli have been studied using a combination of mathematical modeling and experimental analysis [1,2]. A majority of these studies has modeled cell signaling at the population level and used population-averaged measurements such as Western blots to infer the dynamics of different proteins in the signaling pathway, as well as the possible network structure of signaling pathways [1]. However, with recent advances in the ability to measure gene and protein expression at the single-cell level (reviewed in [2,3]), it has become possible to analyze signaling dynamics at the single-cell level. In contrast to the observations from population-average studies, the single-cell studies have demonstrated that individual cells in a clonal population may respond differently to the same stimulus, and the population level measurements could mask the temporal dynamics of individual cells [2]. This variability in the responses of individual cells poses a challenge to their implementation in biology and medicine [4]. Therefore, it is important to understand the stochasticity and heterogeneity in the single-cell responses that might be missed in population-averaged measurements.

Advances in experimental tools for single-cell analysis have led to a significant increase in single-cell studies [2,3]. Despite these advancements, it is still difficult to study the single-cell signaling dynamics due to complex interactions at multiple levels between different proteins that are involved in signal transduction [1]. Computational modeling has been proposed as a complementary approach to overcome some of these limitations and gain insights that cannot be obtained solely through experiments [1,2]. A viable and computationally efficient approach to study the cell-to-cell variability is to use a deterministic model with parameters that have distributions [2,5–7]. In this approach, the computational cost is generally reduced by simulating the signaling dynamics through a deterministic modeling approach while the stochasticity is preserved by assigning a set of different parameter values for each simulation based on predetermined parameter distributions.

In order to construct such models, an experimentally validated deterministic model, which can capture average signaling dynamics at the single-cell level, is required. Although various deterministic models have been proposed for several well-studied signal transduction pathways [1,8], many demonstrate good qualitative, but not quantitative, agreement with the experimental data. This has been attributed to, among other factors, the limited breadth of data used for training the model (e.g., models trained using one dataset with a single stimulus concentration), which makes the models unable to make robust predictions under different conditions. Moreover, the identifiability issue of model parameters [9], which arises due to the model structure as well as the limited availability of experimental data of intracellular proteins [10,11], is not always addressed, which may lead to a suboptimal estimation of model parameters [10,12]. Additionally, many models have been constructed and validated based on experimental data obtained from the population-averaged measurements, which mask the signaling dynamics at the single-cell level [2,13,14]. Consequently, these models are inadequate to predict the average signaling dynamics of single cells.

Motivated by the above considerations, we developed a deterministic model that can accurately predict the average signaling dynamics of single cells. We chose lipopolysaccharide (LPS)-induced nuclear factor κB (NFκB) signaling in mouse macrophages for our model system as it is an extensively studied and characterized signaling pathway [8,15,16]. In order to address the issues discussed above, both computational and experimental approaches have been implemented. First, a rigorous numerical scheme is used to identify the most important parameters that are to be estimated in the parameter estimation [17]. Specifically, the sensitivity analysis and the parameter selection method quantitatively assess the significance of each model parameter with respect to experimental measurements under different LPS concentrations and select parameters whose values could be uniquely estimated [10,18]. Second, flow cytometry with intracellular staining is used to measure the average single-cell dynamics of key molecules involved in the NFκB signaling pathway in response to a broad range of LPS concentrations [19,20]. In this study, the intracellular concentrations of the inhibitor of κB-α (IκBα) and tumor necrosis factor α (TNFα) were measured. IκBα is an inhibitor of NFκB activity, and therefore, the IκBα dynamics are inversely correlated with the NFκB dynamics. At the same time, the activated NFκB induces the transcription and translation of TNFα upon the stimulation of LPS; hence, the TNFα can also be used to infer the dynamics of the NFκB signaling pathway [16]. The obtained average single-cell kinetics is used to quantitatively calibrate and validate the model. Third, the discrepancy between the experimental measurements and the model predictions reveals important, yet unconsidered mechanisms, which is validated experimentally afterwards and leads to the model refinement. Through this integrated model development methodology, predictions from the resultant model quantitatively agree with the experimental measurements. Therefore, the proposed model represents a first step towards the construction of single-cell semi-stochastic models to investigate the stochasticity of intracellular NFκB signaling in macrophages.

2. Material and Methods

2.1. Materials and Cell Culture

RAW264.7 cells were obtained from ATCC (Manassas, VA, USA). Dulbecco's Modified Eagle Medium (DMEM) and penicillin/streptomycin were obtained from Invitrogen (Carlsbad, CA, USA). Bovine serum and fetal bovine serum (FBS) were obtained from Atlanta Biologicals (Flowery Branch, GA, USA). Ultrapure LPS derived from *S. minnesota* was obtained from Invivogen (San Diego, CA, USA). RAW264.7 macrophages were cultured in DMEM supplemented with 10% FBS, penicillin (200 U/mL) and streptomycin (200 μg/mL) at 37 °C in a 5% CO_2 environment.

2.2. Flow Cytometry Analysis

The expression of TNFα and IκBα under different experimental conditions was determined using flow cytometry. RAW264.7 cells were seeded into round-bottomed 96-well plate and stimulated with different concentrations of LPS for the indicated time. Golgiplug™ (BD Biosciences, San Jose, CA, USA) was added along with LPS for TNFα detection experiments to block secretion of TNFα. Cells were then stained with Alexa Flour 700 fluorescence-tagged TNFα antibody (BD Biosciences) and PE-conjugated IκBα antibody (Cell Signaling Technology, Danvers, MA, USA) using the manufacturer's suggested protocol. Stained cells were analyzed using a BD Fortessa flow cytometer (BD Biosciences) at the Texas A&M Health Science Center College of Medicine Cell Analysis Facility. Ten thousands events per sample were acquired, and the data were analyzed using FlowJo software (Tree Star, OR, USA). Cells were gated based on side scattered light (SSC) and forward scattered light (FSC) values to eliminate cell debris, and TNFα- and IκBα-positive cells were gated based on the antibody isotype (see Supplementary Materials Figures S1–S3). All experiments were repeated using at least three different cultures.

2.3. Model Development

The schematic diagram of the NFκB signaling pathway is illustrated in Figure 1. The model used in this study was adopted from Caldwell et al. [21], which takes the extracellular LPS concentration as an input to predict the kinetics of key biomolecules in the NFκB signaling pathway. In this model, by forming a complex with Toll-like receptor 4 (TLR4), LPS activates IκB kinase (IKK) through myeloid differentiation primary response 88 (MyD88)- or TIR (Toll/Interleukin-1 receptor)-domain-containing adaptor-inducing interferon-β (TRIF)-dependent activation of TNF receptor-associated factor 6 (TRAF6). The activated IKK in turn promotes the translocation of NFκB to the nucleus, where the nuclear NFκB induces the transcription of NFκB inhibitors (IκB-α, -β, -ε and A20), as well as TNFα. Once translated, these inhibitors inhibit the NFκB signaling pathway. In contrast, the translated TNFα is secreted to the extracellular medium, and some of the secreted TNFα proteins will bind with TNFα receptor (TNFR) on the cellular membrane to initiate the TNFα-induced NFκB signaling pathway (see [21–23] for details of the model).

Additionally, nonlinear functions proposed by Junkin et al. [24] were added to describe how the rates of TNFα production and secretion increase as the amount of activated TRIF complex increases. This model incorporates the TLR4-mediated NFκB dynamics induced by LPS, as well as the production of TNFα in macrophages (see [21,23] for details). For the purpose of this study, two modifications were made to the model presented by Caldwell et al. [21]. First, transcription delays were ignored to facilitate the simplicity of subsequent calculations for sensitivity analysis and parameter estimation. Second, a new role of A20 protein, which was introduced in the previous model [21,23] as an inhibitor of the TNFα-induced NFκB signaling [23,25,26], was included in the modified model to downregulate the LPS-induced signaling through deubiquitinating of TRAF6 [27].

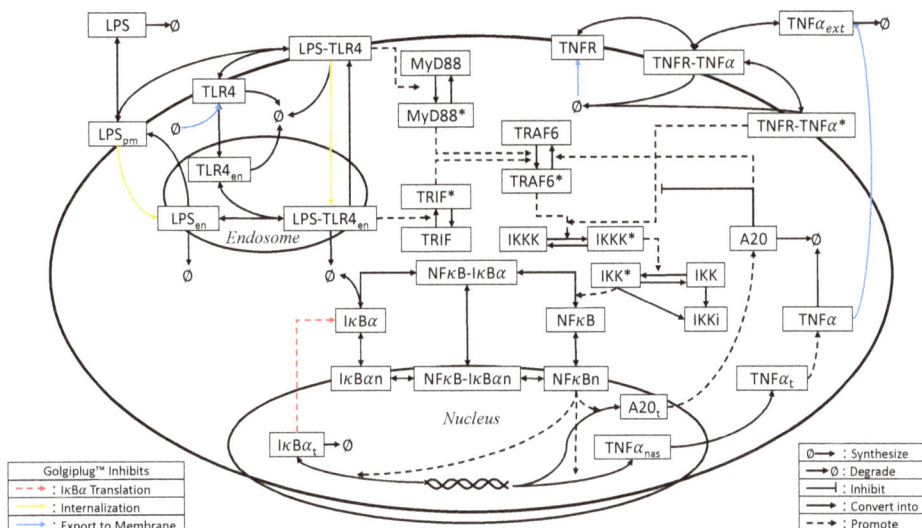

Figure 1. Schematic diagram for the LPS-NFκB-TNFα signaling pathway. Due to space limitation, TRIF-dependent regulation of TNFα production, IκBβ and IκBε-dependent NFκB deactivation and eIF2α-induced translation inhibition are not illustrated. Furthermore, some states related to TNFα-induced activation of IKK kinase (IKKK) are not shown. Colored arrows indicate the processes affected by the addition of Golgiplug™ (see the text for details).

For this study, the TNFα production at the single-cell level was measured using flow cytometry by adding Golgiplug™ since brefeldin A, the active agent of Golgiplug™, causes the Golgi apparatus to merge with endoplasmic reticulum (ER) and inhibits protein export from the Golgi complex [28,29]. Hence, the addition of Golgiplug™ enabled us to measure average single-cell production of TNFα. On the other hand, because Golgiplug™ interferes with the normal cellular processes, it inevitably affects the NFκB signaling dynamics. Specifically, Golgiplug™ suppresses the expression of receptors on the cellular membrane, which negatively regulates the LPS-mediated NFκB signaling pathway in different ways. First, the addition of Golgiplug™ can block the translocation of TLR4 and its accessory molecules from the Golgi complex, which leads to the termination of signaling as these receptors are not replenished after turnover [28,30–32]. Similarly, TNFR is also depleted from the cellular membrane due to Golgiplug™ [33,34], which may inhibit subsequent TNFα autocrine and paracrine signaling [35–37]. Second, Golgiplug™ can hinder the membrane expression of the cluster of differentiation 14 (CD14), which regulates the endocytosis of LPS or the TLR4-LPS complex [38–40]. Therefore, the TRIF-dependent pathway, which is initiated only after LPS or LPS-TLR4 is endocytosed into cytoplasm [5,41], can also be partially impaired. Lastly, the secretion of TNFα proteins translated in response to the NFκB activation will also be inhibited, which helps measure the TNFα production at the single-cell level.

Consequently, the dynamic effects of Golgiplug™ were parameterized and included in the model by the following equations:

$$G = \frac{t}{t + \tau}$$

$$k_{S_{TNFR},m} = k_{S_{TNFR}}(1 - G)$$

$$k_{S_{TLR4},m} = k_{S_{TLR4}}(1 - G)$$

$$k_{en_{LPS},m} = k_{en_{LPS}}(1 - G) \tag{1}$$

$$k_{en_{cp},m} = k_{en_{cp}}(1 - G)$$

$$k_{sec,m} = k_{sec}(1 - G)$$

where G is the normalized activity of Golgiplug™, t is the elapsed time from the addition of Golgiplug™, τ is the characteristic time associated with Golgiplug™ activity, $k_{S_{TNFR}}$ and $k_{S_{TLR4}}$ are the constitutive synthesis rates of TNFR and TLR4, respectively, in the absence of Golgiplug™, $k_{en_{LPS}}$ and $k_{en_{cp}}$ are the endocytosis rates of LPS and the LPS-TLR4 complex, respectively, in the absence of Golgiplug™, k_{sec} is the TNFα secretion rate in the absence of Golgiplug™ and $k_{S_{TNFR},m}$, $k_{S_{TLR4},m}$, $k_{en_{LPS},m}$, $k_{en_{cp},m}$ and $k_{sec,m}$ are the corresponding rates in the presence of Golgiplug™. After Golgiplug™ is added to the cells at $t = 0$, G slowly increases from zero to one, which corresponds to no inhibition of protein export to complete inhibition of protein export from the Golgi apparatus in the presence of Golgiplug™.

Since the signaling kinetics under the stimulation of LPS in the presence of Golgiplug™ were measured experimentally, the dynamic model that consists of the model presented in [21] and Equation (1) was used to simulate the dynamics of LPS-induced NFκB signaling in the presence of Golgiplug™. In general, the dynamic model that simulates the signaling pathway can be represented by a set of nonlinear ordinary differential equations as follows:

$$\frac{dx}{dt} = f(x, \theta; u)$$

$$y = g(x, \theta; u) \tag{2}$$

where x represents the concentration of the biomolecules involved in the signaling pathway (i.e., a vector of states), θ is a vector of model parameters that describe the biochemical reaction rates in the process, u is the concentration of LPS added to the cells (i.e., the process input), and y is the model output (i.e., the experimental measurements predicted by the model). When Golgiplug™ is added, Equation (1) is included in Equation (2), and the overall model consists of 49 states and 146 parameters (see Supplementary Materials Tables S1-S2 and Equations (S1)–(S60)).

2.4. Parameter Estimation

Since we added the Golgiplug™ module to the model developed by Caldwell et al. [21], the integrated dynamic model (the model presented in [21] and Equation (1)) was quantitatively calibrated by estimating its parameters using experimental measurements in response to different LPS concentrations in the presence of Golgiplug™.

The model parameter values were estimated by minimizing the difference between the experimental measurements and the model predictions of the protein concentration. In this work, we used flow cytometry to measure two key molecules in the LPS-induced NFκB signaling pathway: TNF α and IκBα. Since flow cytometry does not provide direct measurements of protein concentration, the mean fluorescence intensity (MFI), which is a measure of the number of copies of the target molecule per cell, was used to infer the protein concentration by assuming a linear relationship between MFI and protein concentration. The experimental data and model prediction were compared based on fold changes of MFI, which are defined as follows:

$$y_{I\kappa Ba}(t) = \frac{(x_{I\kappa Ba}(t) + x_{I\kappa Ba_n}(t) + x_{NF\kappa B\text{-}I\kappa Ba}(t) + x_{NF\kappa B\text{-}I\kappa Ba_n}(t))}{(x_{I\kappa Ba,0} + x_{I\kappa Ba_n,0} + x_{NF\kappa B\text{-}I\kappa Ba,0} + x_{NF\kappa B\text{-}I\kappa Ba_n,0})} \approx \frac{I_{I\kappa Ba}(t) - I_{I\kappa Ba,c}}{I_{I\kappa Ba,0} - I_{I\kappa Ba,c}}$$
$$y_{TNFa}(t) = \frac{x_{TNFa}(t)}{x_{TNFa,0}} \approx \frac{I_{TNFa}(t) - I_{TNFa,c}}{I_{TNFa,0} - I_{TNFa,c}} \tag{3}$$

where $y_{I\kappa Ba}(t)$ and $y_{TNFa}(t)$ are the fold changes of the IκBa and TNFα concentration at time t, $x_{I\kappa Ba}$, $x_{I\kappa Ba_n}$, $x_{NF\kappa B\text{-}I\kappa Ba}$, $x_{NF\kappa B\text{-}I\kappa Ba_n}$ and x_{TNFa} are the cytoplasmic IκBa, nuclear IκBa, cytoplasmic IκBa-NFκB complex, nuclear IκBa-NFκB complex and intracellular TNFα concentration, respectively, $x_{i,0}$ is the initial concentration of the corresponding biomolecules, $I_{I\kappa Ba}$ and I_{TNFa} are the MFI of IκBa and intracellular TNFα, respectively, and $I_{j,0}$ and $I_{j,c}$, $\forall j = \{I\kappa Ba, TNFa\}$, are the corresponding MFI at $t = 0$ and MFI of negative control, respectively. In each cell, IκBa can be part of four biomolecules ($x_{I\kappa Ba}$, $x_{I\kappa Ba_n}$, $x_{NF\kappa B\text{-}I\kappa Ba}$, $x_{NF\kappa B\text{-}I\kappa Ba_n}$); however, flow cytometry measurements can only provide the total IκBa concentration in each cell. Therefore, the simulated concentrations of four IκBa-containing biomolecules were initially summed, and the fold change of the sum (i.e., $y_{I\kappa Ba}$) was computed to compare with the measurements in the subsequent parameter estimation procedure.

One of the biggest challenges in estimating parameters of signaling pathways with a large number of parameters is the parameter identifiability issue [10]. That is, the exact values of some model parameters cannot be uniquely determined from experimental measurements even if a large amount of experimental measurements are available [10,11]. As the proposed model has a large number of parameters, not all the model parameters can be estimated. To this end, a subset of the model parameters, which can be uniquely estimated from the available experimental measurements, was identified through a parameter selection method [10,18]. Only these parameters were estimated against the experimental data.

First, local sensitivity analysis [10,42] was performed to compute two different sensitivity matrices $\mathbf{S}_1$ and $\mathbf{S}_2$ to quantify the effect of each model parameter on $y_{I\kappa Ba}$ and y_{TNFa} (i.e., the process outputs). $\mathbf{S}_1$ and $\mathbf{S}_2$ represent the sensitivity matrices of the model parameters with respect to $y_{I\kappa Ba}$ and y_{TNFa}, respectively, when the cells were stimulated with LPS in the presence of Golgiplug^{TM}. Specifically, a sensitivity matrix is defined as:

$$\mathbf{S}_i = \begin{bmatrix} \frac{\partial y_i(t_1)}{\partial \theta_1} & \cdots & \frac{\partial y_i(t_1)}{\partial \theta_{n_p}} \\ \vdots & \ddots & \vdots \\ \frac{\partial y_i(t_{N_t})}{\partial \theta_1} & \cdots & \frac{\partial y_i(t_{N_t})}{\partial \theta_{n_p}} \end{bmatrix}, \qquad \forall i = \{I\kappa Ba, TNFa\} \tag{4}$$

where n_p is the number of parameters in θ in Equation (2), and $\partial y_i(t_l)/\partial \theta_j$ quantifies the effect of a parameter θ_j on an output y_i at $t = t_l, \forall l = 1, \cdots, N_t$, where N_t is the number of measurement instants. $\partial y_i(t_l)/\partial \theta_j$ can be computed by the following equation:

$$\frac{\partial y_i(t_l)}{\partial \theta_j} = \frac{\partial g_i(t_l)}{\partial x^T} \frac{\partial x}{\partial \theta_j} + \frac{\partial g_i(t_l)}{\partial \theta_j} \tag{5}$$

Additionally, the term $\partial x/\partial \theta_j$ in Equation (5) can be computed by integrating the following equation along with Equation (2):

$$\frac{d}{dt} \frac{\partial x(t_l)}{\partial \theta_j} = \frac{\partial f(t_l)}{\partial x^T} \frac{\partial x}{\partial \theta_j} + \frac{\partial f(t_l)}{\partial \theta_j} \tag{6}$$

Second, the Gram–Schmidt orthogonalization method [10,18] was used to identify the p_i most important model parameters to be estimated for each $\mathbf{S}_i, \forall i = 1, 2$. Here, p_i is the number of singular values of $\mathbf{S}_i$ whose magnitudes are at least 5% of the largest singular value [17,18]. As a result, the parameter subset to be estimated, $\theta_s \in \mathbb{R}^{p \times 1}$ where $p \leq p_1 + p_2$, is chosen as the union of the selected parameters from $\mathbf{S}_1$ and $\mathbf{S}_2$. Third, the least-squares problem was solved to estimate the values of θ_s

by minimizing the difference between the model predictions and the experimental data of $y_{TNF\alpha}$ and $y_{I\kappa B\alpha}$ while the values of the unselected parameters were fixed at their nominal values selected from the literature [21,24,43,44] with some modifications.

In this study, three LPS concentrations (10, 50 and 250 ng/mL) were used to stimulate cells, and the MFI of IκBα and TNFα were measured at $t = 0, 10, 20, 30, 60, 120, 240$ and 360 min after the addition of LPS with Golgiplug^{TM} (i.e., t_l, $\forall\, l = 1, \cdots, 7$). Specifically, the MFI data from 10 and 250 ng/mL of LPS (i.e., u_k, $\forall\, k = 1, 2$) were used to estimate the parameter values, while the dataset from 50 ng/mL LPS was used to validate the model with the updated parameters. Then, the least-squares problem is formulated as follows:

$$\min_{\theta_s} \sum_{k=1}^{2} \sum_{l=1}^{7} \left[\left(\frac{y_{I\kappa B\alpha,k,1}(t_l) - \hat{y}_{I\kappa B\alpha,k,1}(t_l)}{\hat{y}_{I\kappa B\alpha,k,1}(t_l)} \right)^2 + \left(\frac{y_{TNF\alpha,k,1}(t_l) - \hat{y}_{TNF\alpha,k,1}(t_l)}{\hat{y}_{TNF\alpha,k,1}(t_l)} \right)^2 \right] \tag{7}$$

$$\text{s.t.} \quad \frac{dx_{k,i}}{dt} = f_i(x_{k,i}, \theta_s; u_k), \quad x_{k,i}(t = 0) = x_0, \quad \forall\, i = 1, 2 \tag{8}$$

$$y_{j,k,1} = g_j(x_{k,i}, \theta_s; u_k), \quad j = \{I\kappa B\alpha, TNF\alpha\} \tag{9}$$

$$x^{lb} \leq x_{k,i} \leq x^{ub} \tag{10}$$

$$\theta_s^{lb} \leq \theta_s \leq \theta_s^{ub} \tag{11}$$

where $y_{I\kappa B\alpha,k,1}(t_l)$ and $y_{TNF\alpha,k,1}(t_l)$ are the simulated fold changes of IκBα and TNFα, respectively, through Equation (9) at $t = t_l$ under the initial LPS concentration of u_k in the presence of Golgiplug^{TM}, $\hat{y}_{I\kappa B\alpha,k,1}$ and $\hat{y}_{TNF\alpha,k,1}$ are the corresponding experimentally measured fold changes and x_0 is the vector of the initial conditions of x (see Supplementary Materials Table S1).

In the least-squares problem of Equations (7)–(11), the objective function of Equation (7) computes the difference between model predictions and the experimental measurements of the proteins in the presence of Golgiplug^{TM}. As a whole, the objective function minimizes the difference by varying the values of θ_s. While Equation (8) is integrated to compute the predicted protein concentration $x_{k,i}$, f_1, which includes Equation (1), is used if Golgiplug^{TM} is present; otherwise, f_2, which does not involve Equation (1), is integrated. The initial condition of the model, $\hat{x}_0$, is assumed based on a previous study [21,23]. Equations (10)–(11) impose lower and upper bounds on the states and parameters, respectively, based on previous studies and underlying biological knowledge [5,21,23].

It should be noted that we preserved one set of the experimental measurements (one obtained under 50 ng/mL of LPS) to validate the parameter estimation results [45]. As Equations (7)–(11) are likely to be non-convex, the choice of the initial guesses is important. In this study, the initial guesses for the above least-squares problem were obtained from Caldwell et al. [21], which were validated experimentally by comparing with the population-level measurements. Therefore, the parameter values estimated by Caldwell et al. [21] were suitable initial guesses. At the same time, Equations (7)–(11) were solved multiple times with different initial values to avoid any suboptimal optima. Model simulations and the parameter estimation were performed in MATLAB via its functions *ode15s* and *fmincon*. The absolute and relative tolerance criterion for ode15s were set as 10^{-9}, and *fmincon* was implemented with *multistart* to obtain a better result by solving Equations (7)–(11) multiple times with different initial conditions.

3. Results

Profiles of *de novo* synthesized intracellular TNFα under the stimulation of LPS in the presence of Golgiplug^{TM} demonstrated that the TNFα production increased around one hour after the stimulation (Figure 2). At around the same time, the IκBα concentration reached its minimum, which is consistent with experimental observations in the literature [46–48]. Subsequently, the IκBα concentration increased due to the induction of IκB transcript (IκB$_t$) by nuclear translation of NFκB, while the TNFα production rate slowed down beyond 4 h of LPS stimulation (Figure 2). It should be noted that no experiments were conducted beyond 6 h after LPS was added to the cell culture based on the manufacturer's guideline on Golgiplug™ use. This is most likely based on the fact that Golgiplug™ might induce the apoptosis of cells exposed to it for a long time [49,50]. As a result, the calibrated model is more suitable to describe the early NFκB signaling pathway ($\leq$6 h) upon the LPS stimulation.

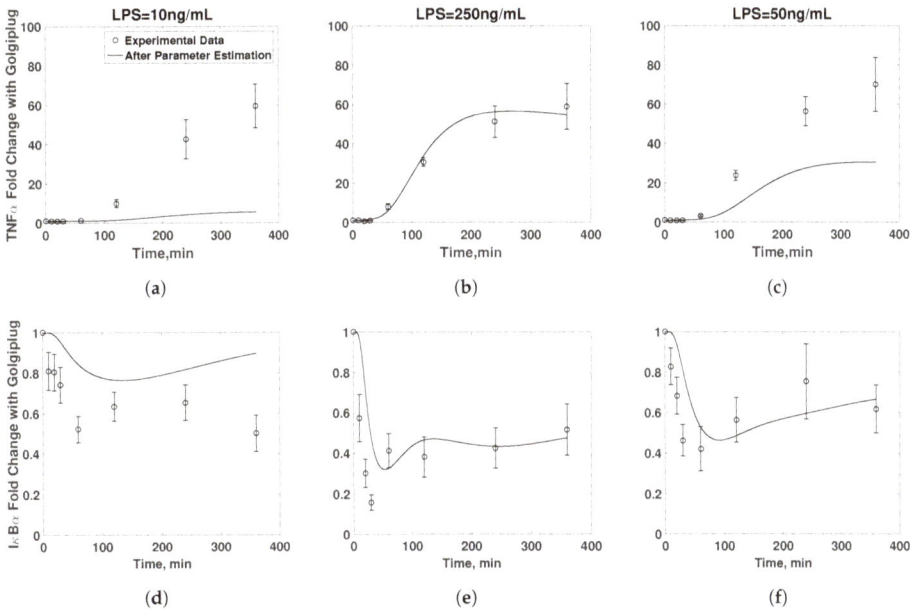

Figure 2. Parameter estimation before considering the Golgiplug™-induced ER stress. (**a**–**c**) Measured (empty circle) and simulated (solid line) fold changes of intracellular TNFα concentrations over time were plotted under different LPS concentrations in the presence of Golgiplug™. (**d**–**f**) Measured (empty circle) and simulated (solid line) fold changes of IκBα concentrations over time were plotted under different LPS concentrations in the presence of Golgiplug™. Indicated amounts of LPS were used for experiments and simulations. Experimental data are given as means ± SEM (standard error of means) with at least *n* = 6.

3.1. Model Validation

Based on the criteria outlined above in the previous section, six parameters (Table 1) were selected for parameter estimation. Figure 2 shows the simulated profiles of intracellular TNFα and IκBα in macrophages under the stimulation of LPS in the presence of Golgiplug™ after the parameter estimation. While the model predictions agreed well with the experimental data obtained for 250 ng/mL of LPS, less concordance was observed between simulations and experimental data for 10 ng/mL of LPS. Specifically, the simulated concentration of intracellular TNFα was one order of magnitude lower than the MFI data, while the simulated IκBα dynamics were qualitatively similar to the measured MFI values. Since the discrepancy between the model prediction and the experimental measurements was pronounced with 10 ng/mL of LPS, we hypothesized that the lack of agreement between the simulations and experimental data was because the effects of Golgiplug™ addition were not adequately represented in the model structure and were more pronounced at the lower LPS concentration.

Table 1. The selected parameters when Golgiplug™-induced IκB translation inhibition was not considered.

Parameter
TLR4 constitutive generation rate
IKKK-mediated IKK activation (IKK → IKKa)
IκBα transcript degradation rate
Hill coefficient of IκBα transcription
Hill coefficient of IκBε transcription
Hill coefficient of TNFα transcription

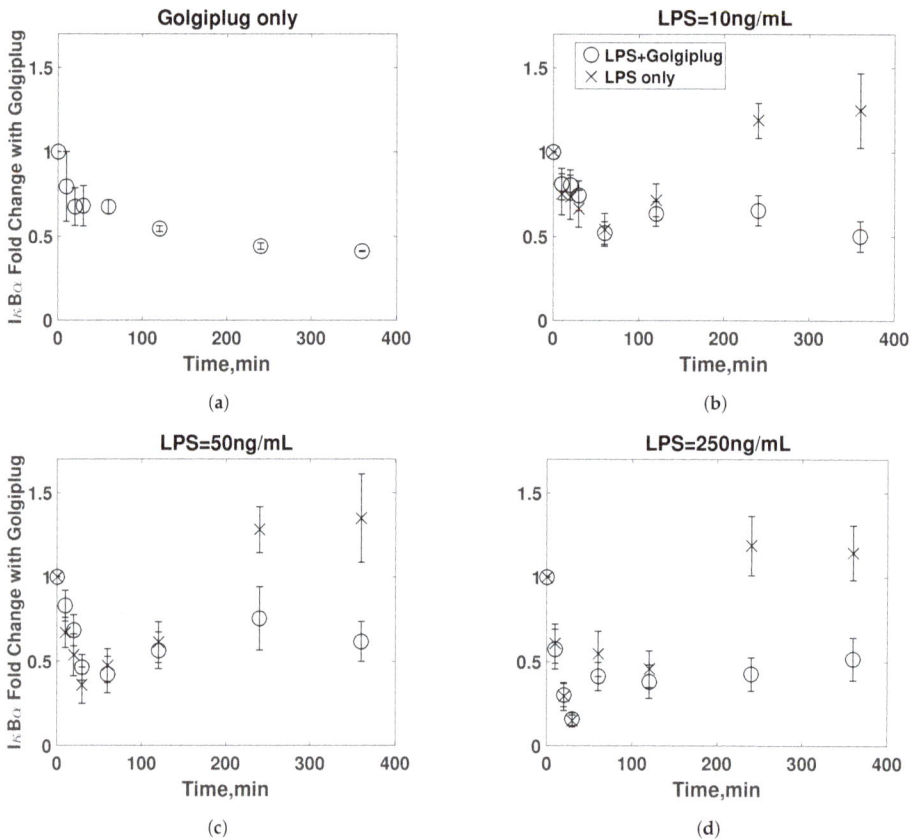

Figure 3. Kinetics of IκBα fold changes when the cells were stimulated by (**a**) 0, (**b**) 10, (**c**) 50, and (**d**) 250 ng/mL of LPS in the presence (empty circles) or absence (x marks) of Golgiplug™. Data are given as means ± SEM with at least *n* = 3.

3.2. Golgiplug™-Induced ER Stress

One possible explanation for this discrepancy could be that the addition of Golgiplug™ induced other signaling pathways, which altered NFκB signaling dynamics [51]. As Golgiplug™ prevents protein secretion by causing collapse of the Golgi apparatus into the ER, synthesized proteins will be redistributed from the Golgi complex into the ER [29]. A direct consequence of Golgiplug™ addition could be accumulation of newly synthesized proteins in ER, which may induce ER stress [51]. It is well established that the ER stress leads to phosphorylation of eukaryotic initiation factor 2 α-subunit (eIF2α), which partially inhibits the translation of IκB in the NFκB signaling pathway [51–54]. This could lead to a decrease in the overall kinetics of the LPS-induced NFκB signaling as the concentration of IκB proteins would be kept lower, leading to the aforementioned mismatch between the model predictions and experimental data. Since the low LPS concentration induces less IκBα and its isomers (IκBβ and IκBε) (Figure 2), the entire LPS-induced NFκB pathway dynamics would be affected more significantly by Golgiplug™ at a lower LPS concentration than at a high LPS concentration if the translation of IκBα and its isomers is partially inhibited. If this is true, it can lead to the pronounced disagreement between the model prediction and the experimental measurement under the stimulation of 10 ng/mL LPS as shown in Figure 2.

Therefore, we examined whether the Golgiplug™ addition could modulate IκB levels in macrophages. First, the fold change in IκBα MFI with the stimulation of LPS alone, Golgiplug™ alone, and LPS and Golgiplug™ in macrophages were compared. Figure 3a shows that Golgiplug™ alone lowers the concentration of IκBα, and Figure 3b–d show that the IκBα kinetics were altered when the cells were stimulated with LPS and Golgiplug™. While IκBα levels initially decreased when cells were exposed to LPS alone, they recovered to pre-stimulation levels after 3 h of exposure. However, when LPS was added along with Golgiplug™, IκBα levels continued to be lower than pre-stimulation levels (Figure 3b–d). These results suggested that Golgiplug™ could affect the IκBα kinetics (presumably through the eIF2α phosphorylation) [52,54]. This also explains the observations in Figure 2a–c, where the intracellular TNFα concentration continued to increase since the Golgiplug™-induced response prolonged the NFκB activation by inhibiting the IκBα synthesis.

3.3. Model Refinement

In order to account for the Golgiplug™-induced translation inhibition, the following equation was considered in addition to Equation (1):

$$k_{tl_i,m} = k_{tl_i} \left(1 - \frac{\nu G}{G + K}\right) \tag{12}$$

where $k_{tl_i,m}$ and k_{tl_i} are the IκBi, $\forall\, i = \alpha, \beta, \epsilon$, translation rates in the presence and absence of Golgiplug™, respectively, ν is a coefficient for the maximum translation inhibition and K is the Michaelis constant of the eIF2α phosphorylation. Equation (12) was included in Equation (2) along with Equation (1) for an accurate simulation. The process affected by this translation inhibition is shown in Figure 1 via a red arrow.

The proposed dynamic model was calibrated again using the parameter estimation procedure as described above. Since the additional measurements of the IκBα dynamics in the absence of Golgiplug™ were obtained, an extra sensitivity matrix was calculated, and the following was added to the objective function (7):

$$\sum_{k=1}^{2} \sum_{l=1}^{7} \left(\frac{y_{I\kappa Ba,k,2}(t_l) - \hat{y}_{I\kappa Ba,k,2}(t_l)}{y_{I\kappa Ba,k,2}(t_l)}\right)^2$$

where $\hat{y}_{I\kappa Ba,k,2}$ and $y_{I\kappa Ba,k,2}$ are the simulated and measured IκBα fold change, respectively, in the absence of Golgiplug™.

3.4. Final Model Validation

Based on the updated model structure and the available experimental data, the aforementioned parameter selection approach determined eight parameters, which could be uniquely estimated (Table 2). Most of the parameters selected by the proposed parameter selection procedure were relevant to the core NFκB-IκB feedback system such as Hill coefficients for IκB-α and -ε transcription, IKK deactivation, and IκBα transcript degradation rate. The remaining identified parameters are the TLR4 constitutive generation rate, C1 (TNFR complex [23]) deactivation rate and eIF2α phosphorylation coefficient, which are most relevant to the LPS- and TNFα-induced NFκB activation, as well as the effects of Golgiplug™, respectively. Hence, all major processes considered in this system, which included the LPS- and TNFα-induced NFκB signaling pathway in the presence of Golgiplug^{TM}, were quantitatively validated against the single-cell experimental data.

Table 2. Selected parameters and their newly estimated values for the final model.

Parameter	New Value
Coefficient for eIF2α phosphorylation (ν)	1.00
A20-mediated C1 deactivation	9.04×10^3 (μM min)$^{-1}$
TLR4 constitutive generation rate	3.75×10^{-2} μM min^{-1}
IKKK-mediated IKK activation	4.75×10^3 (μM min)$^{-1}$
Constitutive inactivation of IKK	2.85×10^{-2} min^{-1}
IκBα mRNA degradation rate	5.83×10^{-3} min^{-1}
Hill coefficient of IκBα transcription	4.16
Hill coefficient of IκBϵ transcription	5.00

Figure 4 shows simulated fold changes in IκBα and intracellular TNFα after parameter estimation. The simulated profiles were again compared with the experimental data. The normalized root-mean-squares of the parameter estimation results before and after the incorporation of the Golgiplug^{TM} model (Equation (12)) were 3.8 and 2.5, respectively, which demonstrated the improvement of the model fidelity. Overall, the model predictions were in qualitative and quantitative agreement with both training datasets and validation datasets, as well as the literature data, which validated the prediction capability of the calibrated model, as well as our hypothesis on the effect of Golgiplug^{TM} on the inhibition of IκB translation. The results demonstrated that the calibrated model is capable of predicting input-output responses in the NFκB pathway. Additionally, the predictions from the current model were compared with the model proposed by Caldwell et al. [21] (Figure 4g–i). The proposed model was able to predict the observed IκBα dynamics under all LPS concentrations more accurately than the previous model by Caldwell et al. [21], which again demonstrated that the predictive capability of the model was improved in terms of simulating the IκBα dynamics.

In order to further assess the predictive capability of the newly calibrated model, the simulated dynamics of nuclear NFκB levels (i.e., activated NFκB) in the absence of Golgiplug^{TM} were computed and plotted in Figure 5a. The maximum NFκB translocation to the nucleus occurred within 2 h of LPS addition, which was consistent with previous experimental studies [15,55,56]. Moreover, as the LPS concentration increased, the nuclear NFκB levels reached their maximum value earlier (i.e., at 50, 60, 75 and 105 min after adding LPS), and the areas under the curves in Figure 5a, which were computed as indicators of the signal strength, were around 20 μM·min for different concentrations of LPS. Interestingly, a 25-fold change in the LPS concentration only resulted in less than a 100% change in the signal strength. This observation was consistent with single-cell studies by Tay et al. [14,57], where they observed a relatively constant peak intensity and decreasing response time of the NFκB signal in mouse 3T3 cells upon TNFα or LPS stimulation.

Figure 5b shows the predicted amount of TNFα secreted under different LPS stimulation conditions in macrophages. As the LPS concentration increased, the concentration of secreted TNFα increased, which was expected since the signal (area under the peak) became stronger. Furthermore, similar to previous studies [15,21,58], the TNFα concentration peaked around 5 h after stimulation and gradually declined thereafter; however, the rate of decline was slower than that reported by Maiti et al. [15] (Figure 5c), where they measured the TNFα secretion dynamics from RAW264.7 macrophages in response to LPS stimulation at the population level. This observation was consistent with the observation reported by Xue et al. [13], who observed using human monocyte-derived macrophages the amount of TNFα secreted to the medium from a single cell in a cell population was less than that from an isolated single-cell at 20 h after the LPS stimulation. This suggested that the simulated dynamics by the proposed model is qualitatively similar to the signaling dynamics of an isolated single-cell instead of population-averaged dynamics, which was expected since the kinetic data obtained under Golgiplug™ were used to train the model.

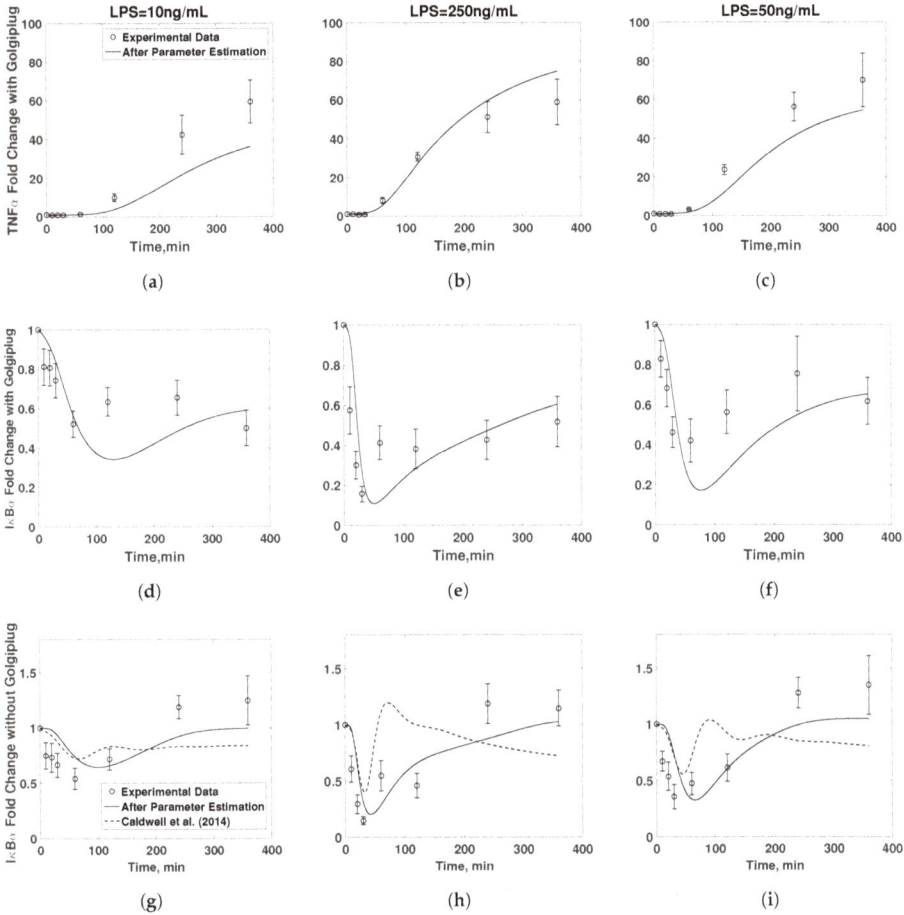

Figure 4. Parameter estimation considering Golgiplug™-induced ER stress. (**a–c**) Measured (empty circle) and simulated (solid line) fold changes of intracellular TNFα concentrations over time were plotted in the presence of Golgiplug™. (**d–f**) Measured (empty circle) and simulated (solid line) fold changes of IκBα concentrations over time were plotted in the presence of Golgiplug™. (**g–i**) Measured (empty circle) and simulated (solid line) fold changes of IκBα concentrations over time were plotted in the absence of Golgiplug™. The IκBα dynamics predicted by the model in [21,24] were also plotted in (**g–i**) for comparison. Indicated amounts of LPS were used for experiments and simulations.

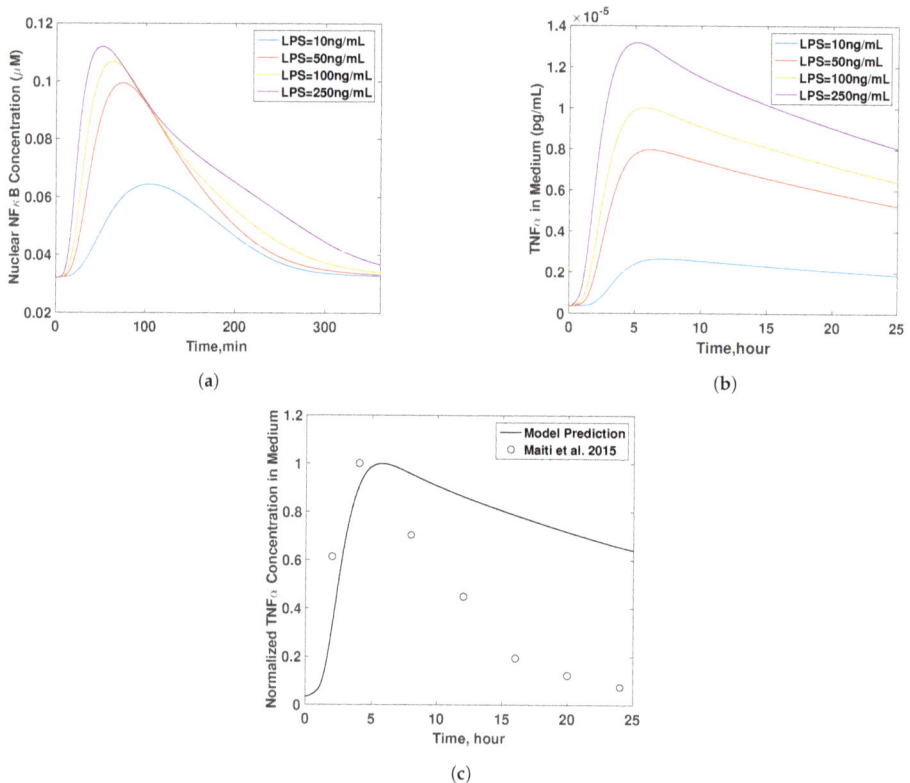

(a)

(b)

(c)

Figure 5. Simulated dynamics of NFκB nuclear translation and TNFα secretion. (**a**) Nuclear NFκB concentration and (**b**) the amount of TNFα secreted to the medium upon stimulation by 10, 50, 100 and 250 ng/mL of LPS. (**c**) The simulated dynamics of TNFα concentration in the medium was compared with the measurement by Maiti et al. [15] in response to 100 ng/mL of LPS. The TNFα concentration at each point was normalized to the maximum value obtained.

4. Discussion

In this study, we have developed a dynamic model that can accurately simulate the average single-cell dynamics of the NFκB signaling pathway by combining the single-cell measurements and a numerical scheme with sensitivity analysis, parameter selection and parameter estimation. The dynamic model was built based on a previously developed NFκB model [21,23,24] and calibrated using the experimental data and the aforementioned numerical scheme. Predictions from the developed dynamic model are in good agreement with the experimental measurements under all LPS concentrations, which demonstrates that the model is capable of simulating the average single-cell dynamics.

Previous studies have used stochastic simulation algorithms such as Gillespie's algorithm [59] and approximate methods of Gillespie's algorithm [60–63] to study single-cell dynamics and investigate heterogeneity in signaling pathways at the single-cell level [2,5,64]. For example, Lipniacki et al. [14,64] proposed a hybrid stochastic-deterministic model of the TNFα-induced NFκB signaling pathway that was able to reproduce the heterogeneous responses observed in the single-cell measurements [14,65] and identify possible origins of the heterogeneity. However, stochastic simulation algorithms are computationally expensive, and they are difficult to fit to experimental measurements for model validation [7,66,67]. A more viable method is a semi-stochastic model, which uses deterministic modeling with model parameters that have distributions [5–7], to reduce the computational cost while still studying the cell-to-cell variability. The dynamic model developed here can accurately simulate average single-cell dynamics and is a first step towards building a semi-stochastic model of the NFκB signaling.

The development of such a deterministic model for building a semi-stochastic model requires accurate parameter estimation, where values of model parameters are estimated by solving an optimization problem (Equations (7)–(11)). However, parameter estimation is a nontrivial problem due to, but not limited to, ill conditioning, over-fitting and the non-identifiability of model parameters [9,68,69]. The ill-conditioning and over-fitting problems during parameter estimation are attributed to the fact that available experimental measurements are usually very limited and noisy, while mathematical models of signaling pathways are often very comprehensive and include a large number of parameters [9,10]. As a result, the solution to the parameter estimation problem is likely to be non-unique or very sensitive to noise present in the experimental measurements. Furthermore, even if a large number of noise-free experimental measurements are available, the value of a parameter cannot be uniquely determined if the parameter is not identifiable [10,11]; hence, it is necessary to check the parameter identifiability a priori.

The model developed in this work contains 148 parameters with limited experimental data, and hence, parameter estimation is very likely to suffer from the aforementioned issues in the parameter estimation procedure. Therefore, we implemented an integrated method combining sensitivity analysis and parameter selection before parameter estimation. Specifically, the sensitivity analysis quantified the effects of each parameter on the measurements, and the parameter selection method selected identifiable parameters via Gram–Schmidt orthogonalization. Then, the values of only the selected parameters were estimated in the parameter estimation, while the values of remaining parameters were fixed at their nominal values, which effectively alleviated the ill-conditioning problem by reducing the degrees of freedom in Equations (7)–(11) [9,10,68].

After parameter estimation, the simulated profiles of intracellular TNFα and IκBα exhibited reasonable agreement between the model predictions and the experimental measurements at all LPS concentrations (Figure 4). Furthermore, as shown in Figure 5c, model predictions after parameter estimation were distinct from that of a cell population as the simulated profiles were closer to the signaling dynamics of isolated single-cells. This was likely because the use of Golgiplug™ inhibited secretion of cytokines [70] and hence minimized potential autocrine and paracrine signaling from the secreted cytokines. This is important as the autocrine and paracrine signaling has been proposed as a key component in determining the overall signaling dynamics of cells in a population [13,35,36,71]. Therefore, the proposed model, which was trained by the single-cell dynamics from flow cytometry in the presence of Golgiplug™, was able to describe the single-cell NFκB dynamics under minimal cytokine feedback.

It should be noted that the current model simulates the LPS-induced NFκB signaling dynamics in a cell, but it does not consider the initiation of the NFκB signaling pathway by TNFα secreted by neighboring cells. Hence, the flow cytometry measurements obtained in the presence of Golgiplug™ are appropriate to identify realistic parameter values to reproduce average single-cell dynamics. At the same time, as flow cytometry measures cellular responses from thousands of cells simultaneously, flow cytometry can provide distributions of the measurements (see Supplementary Materials Figures S1–S3). Based on this statistical information, one can estimate the distributions of the parameters by different methods such as Bayesian approaches [6,7] or generalized polynomial chaos [72]. The model with the estimated parameter distributions is then the semi-stochastic model that can be used to study the heterogeneity in cellular responses.

The present study also suggests that cytokine production data acquired using flow cytometry in the presence of Golgiplug™ should be interpreted cautiously. As Golgiplug™ can block cytokine secretion, it is often used to assess the cytokine production at the single-cell level using flow cytometry [73–75]. The data shown in the work suggest that the dynamics of transcription factors and other signaling intermediates may be altered by the addition of Golgiplug™ (Figure 3).

Therefore, data from studies using Golgiplug™ need to be interpreted cautiously, and a model-based approach like the one presented here can be useful in eliminating the effects of Golgiplug™ and extract true signaling dynamics from flow cytometry data.

5. Conclusions

We systemically extracted the average single-cell dynamics of the LPS-induced NFκB signaling pathway through the integration of sensitivity analysis and a parameter selection scheme with flow cytometry data of key protein intermediates. Based on the measurements and the model structure, key model parameters were identified and estimated to maximize the prediction accuracy of the calibrated model while avoiding overfitting. The mismatch between the model predictions and experimental observations even after the parameter estimation revealed the existence of a previously unconsidered, yet important, mechanism related to Golgiplug™, which was subsequently validated by experiments and led to the update of the proposed model. Then, the resultant model was validated, and the simulated profiles from the updated model were in good agreement with experimental datasets under three different LPS concentrations. This model can be used as the nominal model to construct a deterministic model that has parameters with distributions and can be used to study the stochasticity in signaling.

Supplementary Materials: The model parameters and equations are available at http://www.mdpi.com/2227-9717/6/3/21/s1. Furthermore, representative histograms of IκBα and TNFα levels measured using flow cytometry are provided in the Supplementary Materials Figures S1–S3.

Acknowledgments: Financial support from Artie McFerrin Department of Chemical Engineering, the Texas A&M Energy Institute, the National Institutes of Health (1R01 AI110642-01) to AJand the Ray Nesbitt Chair endowment is acknowledged. We thank Zhang Cheng and Professor Alexander Hoffmann (University of California, Los Angeles) for sharing their MATLAB scripts for simulating the NFκB signaling pathway, and we also thank Professor Juergen Hahn (Rensselaer Polytechnic Institute) for critical comments and suggestions.

Author Contributions: D.L., Y.D., A.J. and J.K. conceived of and designed the study. Y.D. performed the experiments. D.L. and J.K. developed the mathematical model and estimated its parameters. D.L., Y.D., A.J. and J.K. analyzed the data. D.L., Y.D., A.J. and J.K. wrote the paper.

Conflicts of Interest: The authors declare no conflict of interest. The funding sponsors had no role in the design of the study; in the collection, analyses or interpretation of the data; in the writing of the manuscript; nor in the decision to publish the results.

Abbreviations

The following abbreviations are used in this manuscript:

LPS	lipopolysaccharide
NFκB	nuclear factor κB
TNFα	tumor necrosis factor α
IκBα	inhibitor of κB, α
TLR4	Toll-like receptor 4
IKK	IκB kinase
MyD88	myeloid differentiation primary response 88
TIR	Toll/interleukin-1 receptor
TRIF	TIR-domain-containing adaptor-inducing interferon-β
TRAF6	TNF receptor-associated factor 6
IKKK	IKK kinase
ER	endoplasmic reticulum
TNFR	TNFα receptor
CD14	cluster of differentiation 14
MFI	mean fluorescence intensity
IκB$_t$	IκB transcript
eIF2α	eukaryotic initiation factor 2 α-subunit
C1	TNFR complex

References

1. Hughey, J.J.; Lee, T.K.; Covert, M.W. Computational modeling of mammalian signaling networks. *Wiley Interdisciplin. Rev. Syst. Biol. Med.* **2010**, *2*, 194–209.
2. Handly, L.N.; Yao, J.; Wollman, R. Signal transduction at the single-cell level: Approaches to study the dynamic nature of signaling networks. *J. Mol. Biol.* **2016**, *428*, 3669–3682.
3. Gaudet, S.; Miller-Jensen, K. Redefining signaling pathways with an expanding single-cell toolbox. *Trends Biotechnol.* **2016**, *34*, 458–469.
4. Cohen, A.A.; Geva-Zatorsky, N.; Eden, E.; Frenkel-Morgenstern, M.; Issaeva, I.; Sigal, A.; Milo, R.; Cohen-Saidon, C.; Liron, Y.; Kam, Z.; et al. Dynamic proteomics of individual cancer cells in response to a drug. *Science* **2008**, *322*, 1511–1516.
5. Cheng, Z.; Taylor, B.; Ourthiague, D.R.; Hoffmann, A. Distinct single-cell signaling characteristics are conferred by the MyD88 and TRIF pathways during TLR4 activation. *Sci. Signal.* **2015**, *8*, ra69, doi:10.1126/scisignal.aaa5208.
6. Hasenauer, J.; Waldherr, S.; Doszczak, M.; Radde, N.; Scheurich, P.; Allgöwer, F. Identification of models of heterogeneous cell populations from population snapshot data. *BMC Bioinform.* **2011**, *12*, 125.
7. Hasenauer, J.; Waldherr, S.; Doszczak, M.; Scheurich, P.; Radde, N.; Allgöwer, F. Analysis of heterogeneous cell populations: A density-based modeling and identification framework. *J. Process Control* **2011**, *21*, 1417–1425.
8. Williams, R.A.; Timmis, J.; Qwarnstrom, E.E. Computational models of the NF-κB signalling pathway. *Computation* **2014**, *2*, 131–158.
9. Gábor, A.; Villaverde, A.F.; Banga, J.R. Parameter identifiability analysis and visualization in large-scale kinetic models of biosystems. *BMC Syst. Biol.* **2017**, *11*, 54.
10. Kravaris, C.; Hahn, J.; Chu, Y. Advances and selected recent developments in state and parameter estimation. *Comput. Chem. Eng.* **2013**, *51*, 111–123.
11. Raue, A.; Kreutz, C.; Maiwald, T.; Bachmann, J.; Schilling, M.; Klingmüller, U.; Timmer, J. Structural and practical identifiability analysis of partially observed dynamical models by exploiting the profile likelihood. *Bioinformatics* **2009**, *25*, 1923–1929.
12. Kravaris, C.; Seinfeld, J.H. Identification of parameters in distributed parameter systems by regularization. *SIAM J. Control Optim.* **1985**, *23*, 217–241.
13. Xue, Q.; Lu, Y.; Eisele, M.R.; Sulistijo, E.S.; Khan, N.; Fan, R.; Miller-Jensen, K. Analysis of single-cell cytokine secretion reveals a role for paracrine signaling in coordinating macrophage responses to TLR4 stimulation. *Sci. Signal.* **2015**, *8*, ra59, doi:10.1126/scisignal.aaa2155.
14. Tay, S.; Hughey, J.J.; Lee, T.K.; Lipniacki, T.; Quake, S.R.; Covert, M.W. Single-cell NF-κB dynamics reveal digital activation and analogue information processing. *Nature* **2010**, *466*, 267.
15. Maiti, S.; Dai, W.; Alaniz, R.C.; Hahn, J.; Jayaraman, A. Mathematical modeling of pro- and anti-inflammatory signaling in macrophages. *Processes* **2015**, *3*, 1–18.
16. Hayden, M.S.; Ghosh, S. NF-κB, the first quarter-century: Remarkable progress and outstanding questions. *Genes Dev.* **2012**, *26*, 203–234.
17. Chu, Y.; Hahn, J. Parameter set selection via clustering of parameters into pairwise indistinguishable groups of parameters. *Ind. Eng. Chem. Res.* **2009**, *48*, 6000–6009.
18. Yao, K.Z.; Shaw, B.M.; Kou, B.; McAuley, K.B. Modeling ethylene/butene copolymerization with multi-site catalysts: Parameter estimability and experimental design. *Polym. React. Eng.* **2003**, *3*, 563–588.
19. Prussin, C. Cytokine flow cytometry: Understanding cytokine biology at the single-cell level. *J. Clin. Immunol.* **1997**, *17*, 195.
20. Schulz, K.R.; Danna, E.A.; Krutzik, P.O.; Nolan, G.P. Single-cell phospho-protein analysis by flow cytometry. *Curr. Protoc. Immunol.* **2012**, 8–17, doi:10.1002/0471142735.im0817s96.
21. Caldwell, A.B.; Cheng, Z.; Vargas, J.D.; Birnbaum, H.A.; Hoffmann, A. Network dynamics determine the autocrine and paracrine signaling fucntions of TNF. *Genes Dev.* **2014**, *28*, 2120–2133.
22. Hoffmann, A.; Levchenko, A.; Scott, M.L.; Baltimore, D. The IκB-NF-κB signaling module: Temporal control and selective gene activation. *Science* **2002**, *298*, 1241–1245.

23. Werner, S.L.; Kearns, J.D.; Zadorozhnaya, V.; Lynch, C.; O'Dea, E.; Boldin, M.P.; Ma, A.; Baltimore, D.; Hoffmann, A. Encoding NF-κB temporal control in response to TNF: Distinct roles for the negative regulators IκBα and A20. *Genes Dev.* **2008**, *22*, 2093–2101.
24. Junkin, M.; Kaestli, A.J.; Cheng, Z.; Jordi, C.; Albayrak, C.; Hoffmann, A.; Tay, S. High-content quantification of single-cell immune dynamics. *Cell Rep.* **2016**, *15*, 411–422.
25. Krikos, A.; Laherty, C.D.; Dixit, V.M. Transcriptional activation of the tumor necrosis factor α-inducible zinc finger protein, a20, is mediated by κB elements. *J. Biol. Chem.* **1992**, *267*.
26. Lee, E.G.; Boone, D.L.; Chai, S.; Libby, S.L.; Chien, M.; Lodolce, J.P.; Ma, A. Failure to regulate TNF-induced NF-κB and cell death responses in A20-deficient mice. *Science* **2000**, *289*, 2350–2354.
27. Boone, D.L.; Turer, E.E.; Lee, E.G.; Ahmad, R.C.; Wheeler, M.T.; Tsui, C.; Hurley, P.; Chien, M.; Chai, S.; Hitotsumatsu, O.; et al. The ubiquitin-modifying enzyme A20 is required for termination of Toll-like receptor responses. *Nat. Immunol.* **2004**, *5*, 1052–1060.
28. Chardin, P.; McCormick, F. Brefeldin A: The advantage of being uncompetitive. *Cell* **1999**, *97*, 153–155.
29. Ward, T.H.; Polishchuk, R.S.; Caplan, S.; Hirschberg, K.; Lippincott-Schwartz, J. Maintenance of Golgi structure and function depends on the integrity of ER export. *J. Cell Biol.* **2001**, *155*, 557–570.
30. Latz, E.; Visintin, A.; Lien, E.; Fitzgerald, K.A.; Monks, B.G.; Kurt-Jones, E.A.; Golenbock, D.T.; Espevik, T. Lipopolysaccharide rapidly traffics to and from the Golgi apparatus with the Toll-like Receptor 4-MD-2-CD14 complex in a process that Is distinct from the initiation of signal transduction. *J. Biol. Chem.* **2002**, *49*, 47834–47843.
31. Liaunardy-Jopeace, A.; Bryant, C.E.; Gay, N.J. The COPII adaptor protein TMED7 is required to initiate and mediate the anterograde trafficking of Toll-like receptor 4 to the plasma membrane. *Sci. Signal.* **2014**, *7*, ra70, doi:10.1126/scisignal.2005275.
32. Wang, D.; Lou, J.; Ouyang, C.; Chen, W.; Liu, Y.; Liu, X.; Cao, X.; Wanga, J.; Lu, L. Ras-related protein Rab10 facilitates TLR4 signaling by promoting replenishment of TLR4 onto the plasma membrane. *Proc. Natl. Acad. Sci. USA* **2010**, *107*, 13806–13811.
33. Jones, S.J.; Ledgerwood, E.C.; Prins, J.B.; Galbraith, J.; Johnson, D.R.; Pober, J.S.; Bradley, J.R. TNF recruits TRADD to the plasma membrane but not the trans-Golgi Network, the principal subcellular location of TNF-R1. *J. Immunol.* **1999**, *162*, 1042–1048.
34. Neznanov, N.; Kondratova, A.; Chumakov, K.M.; Angres, B.; Zhumabayeva, B.; Agol, V.I.; Gudkov, A.V. Poliovirus protein 3A inhibits tumor necrosis factor (TNF)-induced apoptosis by eliminating the TNF receptor from the cell surface. *J. Virol.* **2001**, *75*, 10409–10420.
35. Xaus, J.; Comalada, M.; Valledor, A.F.; Lloberas, J.; López-Soriano, F.; Argilés, J.M.; Bogdan, C.; Celada, A. LPS induces apoptosis in macrophages mostly through the autocrine production of TNF-α. *Blood* **2000**, *95*, 3823–3831.
36. Covert, M.W.; Leung, T.H.; Gaston, J.E.; Baltimore, D. Achieving stability of lipopolysaccharide-induced NF-κB activation. *Science* **2005**, *309*, 1854–1857.
37. Lombardo, E.; Alvarez-Barrientos, A.; Maroto, B.; Boscà, L.; Knaus, U.G. TLR4-mediated survival of macrophages is MyD88 dependent and requires TNF-α autocrine signalling. *J. Immunol.* **2007**, *178*, 3731–3739.
38. Zanoni, I.; Ostuni, R.; Marek, L.R.; Barresi, S.; Barbalat, R.; Barton, G.M.; Granucci, F.; Kagan, J.C. CD14 controls the LPS-induced endocytosis of Toll-like Receptor 4. *Cell* **2011**, *147*, 868–880.
39. Tan, Y.; Zanoni, I.; Cullen, T.W.; Goodman, A.L.; Kagan, J.C. Mechanisms of Toll-like receptor 4 endocytosis reveal a common immune-evasion strategy used by pathogenic and commensal bacteria. *Immunity* **2015**, *43*, 909–922.
40. Rajaiah, R.; Perkins, D.J.; Ireland, D.D.C.; Vogel, S.N. CD14 dependence of TLR4 endocytosis and TRIF signaling displays ligand specificity and is dissociable in endotoxin tolerance. *Proc. Natl. Acad. Sci. USA* **2013**, *112*, 8391–8396.
41. Kagan, J.C.; Su, T.; Horng, T.; Chow, A.; Akira, S.; Medzhitov, R. TRAM couples endocytosis of Toll-like receptor 4 to the induction of interferon-β. *Nat. Immunol.* **2008**, *9*, 361–368.
42. Chu, Y.; Jayaraman, A.; Hahn, J. Parameter sensitivity analysis of IL-6 signaling pathways. *IET Syst. Biol.* **2007**, *1*, 342–352.
43. Lin, S.C.; Lo, Y.C.; Wu, H. Helical assembly in the MyD88-IRAK4-IRAK2 complex in TLR/IL-1R signalling. *Nature* **2010**, *465*, 885–890.

44. Bagnall, J.; Boddington, C.; Boyd, J.; Brignall, R.; Rowe, W.; Jones, N.A.; Schmidt, L.; Spiller, D.G.; White, M.R.; Paszek, P. Quantitative dynamic imaging of immune cell signalling using lentiviral gene transfer. *Integr. Biol.* **2015**, *7*, 713–725.

45. Moya, C.; Huang, Z.; Cheng, P.; Jayaraman, A.; Hahn, J. Investigation of IL-6 and IL-10 signalling via mathematical modelling. *IET Syst. Biol.* **2011**, *5*, 15–26.

46. Sung, M.H.; Li, N.; Lao, Q.; Gottschalk, R.A.; Hager, G.L.; Fraser, I.D.C. Switching of the relative dominance between feedback mechanisms in lipopolysaccharide-Induced NF-*κ*B signaling. *Sci. Signal.* **2014**, *7*, ra6, doi:10.1126/scisignal.2004764.

47. Tsukamoto, H.; Fukudome, K.; Takao, S.; Tsuneyoshi, N.; Kimoto, M. Lipopolysaccharide-binding protein-mediated Toll-like receptor 4 dimerization enables rapid signal transduction against lipopolysaccharide stimulation on membrane-associated CD14-expressing cells. *Int. Immunol.* **2010**, *22*, 271–280.

48. Sakai, J.; Cammarota, E.; Wright, J.A.; Cicuta, P.; Gottschalk, R.A.; Li, N.; Fraser, I.D.C.; Bryant, C.E. Lipopolysaccharide-induced NF-*κ*B nuclear translocation is primarily dependent on MyD88, but TNF*α* expression requires TRIF and MyD88. *Sci. Rep.* **2017**, *7*, 1428.

49. Shao, R.G.; Shimizu, T.; Pommier, Y. Brefeldin A Is a potent inducer of apoptosis in human cancer cells independently of p53. *Exp. Cell Res.* **1996**, *227*, 190–196.

50. Moon, J.L.; Kim, S.Y.; Shin, S.W.; Park, J.W. Regulation of brefeldin A-induced ER stress and apoptosis by mitochondrial NADP+-dependent isocitrate dehydrogenase. *Biochem. Biophys. Res. Commun.* **2012**, *417*, 760–764.

51. Cláudio, N.; Dalet, A.; Gatti, E.; Pierre, P. Mapping the crossroads of immune activation and cellular stress response pathways. *EMBO J.* **2013**, *32*, 1214–1224.

52. Mellor, H.; Kimball, S.R.; Jefferson, L.S. Brefeldin A inhibits protein synthesis through the phosphorylation of the *α*-subunit of eukaryotic initiation factor-2. *FEBS Lett.* **1994**, *350*, 143–146.

53. Tam, A.B.; Mercado, E.L.; Hoffmann, A.; Niwa, M. ER stress activates NF-*κ*B by integrating functions of basal IKK activity, IRE1 and PERK. *PLoS ONE* **2012**, *7*, e45078.

54. Fishman, P.E.; Curran, P.K. Brefeldin A inhibits protein synthesis in cultured cells. *FEBS Lett.* **1992**, *314*, 371–374.

55. Ando, Y.; Oku, T.; Tsuji, T. Platelet supernatant suppresses LPS-induced nitric oxide production from macrophages accompanied by inhibition of NF-*κ*B signaling and increased Arginase-1 expression. *PLoS ONE* **2016**, doi:10.1371/journal.pone.0162208.

56. Selimkhanov, J.; Taylor, B.; Yao, J.; Pilko, A.; Albeck, J.; Hoffmann, A.; Tsimring, L.; Wollman, R. Accurate information transmission through dynamic biochemical signaling networks. *Science* **2014**, *346*, 1370–1373.

57. Kellogg, R.A.; Tian, C.; Lipniacki, T.; Quake, S.R.; Tay, S. Digital signaling decouples activation probability and population heterogeneity. *eLife* **2015**, *4*, e08931.

58. Noman, A.S.M.; Koide, N.; Hassan, F.; I.-E.-Khuda, I.; Dagvadorj, J.; Tumurkhuu, G.; Islam, S.; Naiki, Y.; Yoshida, T.; Yokochi, T. Thalidomide inhibits lipopolysaccharide-induced tumor necrosis factor-a production via down-regulation of MyD88 expression. *Innate Immun.* **2009**, *15*, 33–41.

59. Gillespie, D.T. Exact stochastic simulation of coupled chemical reactions. *J. Phys. Chem.* **1977**, *82*, 2340–2361.

60. Haseltine, E.L.; Rawlings, J.B. Approximate simulation of coupled fast and slow reactions for stochastic chemical kinetics. *J. Chem. Phys.* **2002**, *117*, 6959–6969.

61. Cao, Y.; Petzold, L.R. The numerical stability of leaping methods for stochastic simulation of chemically reacting systems. *J. Chem. Phys.* **2004**, *121*, 12169–12178.

62. Kwon, J.S.I.; Nayhouse, M.; Christofides, P.D.; Orkoulas, G. Modeling and control of protein crystal shape and size in batch crystallization. *AIChE J.* **2013**, *59*, 2317–2327.

63. Kwon, J.S.; Nayhouse, M.; Christofides, P.D.; Orkoulas, G. Modeling and control of shape distribution of protein crystal aggregates. *Chem. Eng. Sci.* **2013**, *104*, 484–497.

64. Lipniacki, T.; Paszek, P.; Brasier, A.R.; Luxon, B.A.; Kimmel, M. Stochastic regulation in early immune response. *Biophys. J.* **2006**, *90*, 725–742.

65. Nelson, D.E.; Ihekwaba, A.E.C.; Elliott, M.; Johnson, J.R.; Gibney, C.A.; Foreman, B.E.G.; Nelson, V.S.; Horton, C.A.; Spiller, D.G.; Edwards, S.W.; et al. Oscillations in TNF*α* signaling control the dynamics of gene expression. *Science* **2004**, *306*, 704–708.

66. Wilkinson, D.J. Stochastic modelling for quantitative description of heterogeneous biological systems. *Nat. Rev. Genet.* **2009**, *10*, 122.

67. Gupta, A.; Rawlings, J.B. Comparison of parameter estimation methods in stochastic chemical kinetic models: Examples in systems biology. *AIChE J.* **2014**, *60*, 1253–1268.

68. Ashyraliyev, M.; Fomekong-Nanfack, Y.; Kaandorp, J.A.; Blom, J.G. Systems biology: Parameter estimation for biochemicalmodels. *FEBS J.* **2009**, *276*, 886–902.

69. Kiparissides, A.; Koutinas, M.; Kontoravdi, C.; Mantalaris, A.; Pistikopoulos, E.N. 'Closing the loop' in biological systems modeling—from the in silico to the in vitro. *Automatica* **2011**, *47*, 1147–1155.

70. Lamoreaux, L.; Roederer, M.; Koup, R. Intracellular cytokine optimization and standard operating procedure. *Nat. Protoc.* **2006**, *1*, 1507–1516.

71. Lee, T.K.; Denny, E.M.; Sanghvi, J.C.; Gaston, J.E.; Maynard, N.D.; Hughey, J.J.; Covert, M.W. A noisy paracrine signal determines the cellular NF-κB response to lipopolysaccharide. *Sci. Signal.* **2009**, *2*, ra65, doi:10.1126/scisignal.2000599.

72. Xiu, D.; Karniadakis, G.E. The Wiener-Askey Polynomial Chaos for Stochastic Differential Equations. *SIAM J. Sci. Comput.* **2006**, *24*, 619–644.

73. Misumi, Y.; Yuko Misumi, K.M.; Takatsuki, A.; Tamura, G.; Ikehara, Y. Novel blockade by brefeldin A of intracellular transport of secretory proteins in cultured rat hepatocytes. *J. Biol. Chem.* **1986**, *261*, 1139–11403.

74. Bueno, C.; Almeida, J.; Alguero, M.; Sánchez, M.; Vaquero, J.; Laso, F.; Miguel, J.S.; Escribano, L.; Orf, A. Flow cytometric analysis of cytokine production by normal human peripheral blood dendritic cells and monocytes: Comparative analysis of different stimuli, secretion-blocking agents and incubation periods. *Cytometry Part A* **2001**, *46*, 33–40.

75. Gottschalk, R.A.; Martins, A.J.; Angermann, B.R.; Dutta, B.; Ng, C.E.; Uderhardt, S.; Tsang, J.S.; Fraser, I.D.C.; Meier-Schellersheim, M.; Germain, R.N. Distinct NF-κB and MAPK activation thresholds uncouple steady-state microbe sensing from anti-pathogen inflammatory responses. *Cell Syst.* **2016**, *2*, 378–390.

processes

MDPI

Article

Elucidating Cellular Population Dynamics by Molecular Density Function Perturbations

Thanneer Malai Perumal [1] and Rudiyanto Gunawan [2,3,*]

[1]   Sage Bionetworks, Seattle, WA 98109, USA; thanneer.perumal@sagebase.org
[2]   Institute for Chemical and Bioengineering, ETH Zurich, Zurich 8093, Switzerland
[3]   Swiss Institute of Bioinformatics, Lausanne 1015, Switzerland
*   Correspondence: rudi.gunawan@chem.ethz.ch; Tel.: +41-44-633-2134

Received: 21 December 2017; Accepted: 18 January 2018; Published: 23 January 2018

Abstract: Studies performed at single-cell resolution have demonstrated the physiological significance of cell-to-cell variability. Various types of mathematical models and systems analyses of biological networks have further been used to gain a better understanding of the sources and regulatory mechanisms of such variability. In this work, we present a novel sensitivity analysis method, called molecular density function perturbation (MDFP), for the dynamical analysis of cellular heterogeneity. The proposed analysis is based on introducing perturbations to the density or distribution function of the cellular state variables at specific time points, and quantifying how such perturbations affect the state distribution at later time points. We applied the MDFP analysis to a model of a signal transduction pathway involving TRAIL (tumor necrosis factor-related apoptosis-inducing ligand)-induced apoptosis in HeLa cells. The MDFP analysis shows that caspase-8 activation regulates the timing of the switch-like increase of cPARP (cleaved poly(ADP-ribose) polymerase), an indicator of apoptosis. Meanwhile, the cell-to-cell variability in the commitment to apoptosis depends on mitochondrial outer membrane permeabilization (MOMP) and events following MOMP, including the release of Smac (second mitochondria-derived activator of caspases) and cytochrome c from mitochondria, the inhibition of XIAP (X-linked inhibitor of apoptosis) by Smac, and the formation of the apoptosome.

Keywords: mathematical modeling; biological networks; sensitivity analysis; programmed cell death; single cell dynamics; cell population

1. Introduction

Advances in single-cell profiling technology and the application of this technology to study biology at single-cell resolution have demonstrated the ubiquity and functional role of cell-to-cell variability in physiological processes, such as programmed cell death (apoptosis) and stem cell differentiation [1–3]. Besides genetic, epigenetic, and environmental factors, the cellular heterogeneity observed in a given cell population has also been attributed to the inherent stochastic dynamics of cellular processes. For example, gene transcriptional processes have been shown to occur in stochastic (random) bursts [4–6]. Many modeling frameworks have been used to capture cellular heterogeneity—for example, by using ensemble models (EM) of ordinary differential equations (ODEs) [7–9], population balance models (PBMs) [10], stochastic ordinary differential equations (SDEs) [11,12], and chemical master equations (CMEs) [13–15]. In these models, the cell-to-cell variability is described by a probability density or distribution function of cell state variables. Systems analyses have also been developed and applied to gain insight into the dynamics of cell state distribution. For example, several types of parameter sensitivity analysis, including SOBOL sensitivity [16], derivative-based global sensitivity measure (DGSM) [17], glocal analysis [18], extended Fourier amplitude sensitivity test (eFAST) [19], and stochastic sensitivity analysis [20–23] have been used to identify the rate-controlling or bottlenecking processes based on dynamic models of cell distribution.

Parameter sensitivity analysis (PSA) is a common systems analysis that is used to elucidate the dependence of system behavior on system parameters [24–27]. In the PSA, we compute sensitivity coefficients whose magnitudes describe how much system states vary with changes in one or a combination of system parameters. A large sensitivity magnitude means that the system behavior strongly depends on changes in the corresponding parameter(s), an indication of a rate-limiting process. In several publications [28–30], we have shown that the traditional PSA derived using static perturbations to system parameters may lead to incorrect conclusions when the rate-limiting process changes with time. For this reason, we have created a new class of sensitivity analysis based on impulse perturbations on parameters and states, called impulse parameter sensitivity analysis (iPSA) and Green's function matrix (GFM) analysis, respectively [28–30]. By introducing impulse perturbations at different times, the new sensitivity analyses are able to reveal not only which processes are rate limiting but also when they become rate limiting.

In this work, we adapted the concept of impulse perturbation-based PSA for the dynamical analysis of cell-to-cell variability. The new sensitivity analysis, called molecular density function perturbation (MDFP), is based on time-varying perturbations to the probability density or distribution function of the cell state variables. The MDFP sensitivity coefficients are defined using distribution distances in order to account for changes in the cell state distribution beyond the first-order moment (i.e., population mean). We applied the MDFP analysis to a model of programmed cell death in HeLa cell populations [31] and identified key regulators in apoptotic decision making.

2. Material and Methods

Sensitivity analysis of dynamic models of cell distribution has received much interest in recent times along with the rise of systems biology and the increasing attention to single-cell analysis. Novel PSA methods have been developed for the CME models of biological networks [20–23]. Here, the sensitivity coefficients describe changes in the mean of cell state distribution caused by infinitesimal (local) perturbations to the parameter values. Methods for global sensitivity analysis have also been adapted for analyzing cell distribution sensitivities, such as sampling-based partial rank correlation coefficient (PRCC) and variance-based eFAST [19]. Despite their differences, the aforementioned sensitivity analyses and the corresponding sensitivity coefficients are based on static or persistent parameter perturbations. As we have demonstrated previously, such analysis is incapable of elucidating any dynamic transitions of the bottlenecking process [28–30].

2.1. Molecular Density Function Perturbation (MDFP) Analysis

In the following, we formulate the molecular density function perturbation (MDFP) analysis. In MDFP, we describe the cell distribution using a probability density function (PDF) denoted by $f_{\mathbf{X}}(\mathbf{x}, t)$, where $\mathbf{x} \in \mathbb{R}^n$ denotes the cell state vector and t denotes time. This description of cell distribution is flexible enough to accommodate mathematical modeling frameworks that are commonly used to simulate cell population dynamics, including EMs, PBMs, SDEs, and CMEs. In biological network models, the cell state is typically defined by the concentrations of biomolecules. By definition, the (n-tuple) integral of the PDF $\int_a^b f_{\mathbf{X}}(\mathbf{x}, t) d\mathbf{x}$ gives the fraction of the cell population at time t whose states (concentrations) satisfy $\mathbf{a} \leq \mathbf{x} \leq \mathbf{b}$. The basic premise of the MDFP analysis is the same as that of the impulse perturbation-based sensitivity analysis, specifically the GFM analysis [28], which is to introduce a perturbation to the cell state at time τ and quantify the effect of this perturbation at a later time t ($t \geq \tau$). However, the MDFP analysis uses perturbations to the PDF of the cell state.

In deriving the MDP sensitivity coefficients, we start with the following relationship:

$$f_{\mathbf{X}}(\mathbf{x}, t) = f_{\mathbf{X}|\mathbf{X}_\tau}(\mathbf{x}, t | \mathbf{x}_\tau, \tau) f_{\mathbf{X}}(\mathbf{x}_\tau, \tau). \tag{1}$$

The PDF $f_{X|X_\tau}(x, t|x_\tau, \tau)$, also known as the transitional probability or the propensity function, gives the conditional PDF of x at time t given that the cell state is x_τ at time τ ($t \geq \tau$). In the following, we consider introducing a mean shift perturbation to the PDF at time τ to give:

$$f_X^{\Delta+j}(\check{x}_\tau, \tau) = f_X(\check{x}_\tau - \delta e_j, \tau), \tag{2}$$

where $\check{x}_\tau$ denotes the perturbed state variables and e_j denotes the j-th column of the identity matrix. Note that the PDF $f_X^{\Delta+j}(\check{x}_\tau, \tau)$ corresponds to the PDF $f_X(x_\tau, \tau)$ with a positive mean shift of δe_j (i.e., $\check{x}_\tau = x_\tau + \delta e_j$). Given the perturbed PDF $f_X^{\Delta+j}(\check{x}_\tau, \tau)$ at time τ and the propensity function $f_{X|X_\tau}(x, t|x_\tau, \tau)$, we can define the perturbed PDF of the cell state at time t, denoted by $f_X^{\Delta+j,\tau}(x, t)$, as follows:

$$f_X^{\Delta+j,\tau}(\check{x}, t) = f_{X|X_\tau}(\check{x}, t|\check{x}_\tau, \tau) f_X^{\Delta+j}(\check{x}_\tau, \tau). \tag{3}$$

Note that the following equality applies:

$$f_X^{\Delta+j,\tau}(\check{x}, \tau) = f_X^{\Delta+j}(\check{x}_\tau, \tau). \tag{4}$$

In the MDFP analysis, we employ a distribution distance metric to quantify the magnitude of PDF changes caused by the perturbations. Several metrics of distribution distance are available, such as the Kullback–Leibler distance (Δ_{KL}), Jeffrey distance (Δ_J), Jensen–Shannon divergence (Δ_{JS}), engineering metric (Δ_E), Kolmogorov–Smirnov distance (Δ_{KS}), and the Cramer–von Mises distance (Δ_{CVM}). The first four of the aforementioned distribution distances are based on the difference between two PDFs, while the last two are based on the difference between the cumulative density functions (CDFs). For the analysis of programmed cell death (below), we used the Cramer–von Mises distance (see Supplementary Material S1 for the mathematical definitions of the other distribution distances):

$$\Delta_{CVM}\left(f^A(z)\big|\big|f^B(z)\right) = \int_{-\infty}^{\infty}\left(F^A(z) - F^B(z)\right)^2 dz, \tag{5}$$

which, in our experience, gives more reliable sensitivity coefficient calculations. The variable $F(z)$ denotes the CDF of the PDF $f(z)$—i.e., $F(z) = \int_{-\infty}^{z} f(y)dy$.

Following the common definition of sensitivity coefficients [32], we compute the MDFP sensitivity coefficients as the ratio of the changes in the PDF of the cell state at time t and the perturbation introduced at time τ. The sensitivity coefficients are evaluated for a particular cell state variable of interest x_i (the i-th element of x) with respect to a perturbation to δe_j on the state variable x_j, as follows:

$$S_{i,j}^{MDFP}(t, \tau) = sign\left(\frac{\Delta\mu_{X_i}(t)}{\Delta\mu_{x_j}(\tau)}\right) \frac{\Delta_{CVM}\left(f_{X_i}^{\Delta+j,\tau}(\check{x}_i, t)\big|\big|f_{X_i}^{\Delta-j,\tau}(\check{x}_i, t)\right)}{\Delta_{CVM}\left(f_{X_j}^{\Delta+j}(\check{x}_j, \tau)\big|\big|f_{X_j}^{\Delta-j}(\check{x}_j, \tau)\right)}, \tag{6}$$

where $sign(\cdot)$ gives the sign of the argument variable and $\Delta\mu_{X_i}(t)$ denotes the change in the mean of the state variable x_i at time t. The function $f_{X_i}^{\Delta+j,\tau}(\check{x}_i, t)$ denotes the marginal PDF of $f_X^{\Delta+j,\tau}(\check{x}, t)$ with the following definition:

$$f_{X_i}^{\Delta+j,\tau}(\check{x}_i, t) = \int f_X^{\Delta+j,\tau}(\check{x}, t)d\check{x}_{\sim i}. \tag{7}$$

The integration in Equation (7) is performed over all state variables $\check{x}$'s except for $\check{x}_i$. Note that the sensitivity coefficient in Equation (6) is motivated by the centered difference approximation [32], where the sensitivity coefficients are computed using positive and negative perturbations to the system.

The definition of the MDFP sensitivity coefficients is analogous to the Green's function matrix (GFM) sensitivity [28,32]. We can visualize $S_{i,j}^{MDFP}(t, \tau)$ using a heatmap as shown in Figure 1. The magnitudes of the sensitivities represent the degree of importance, while the signs of the sensitivity coefficients reflect the direction of the mean change. A positive sensitivity coefficient indicates that the mean change of x_i at time t is in the same direction as the mean shift perturbation to x_j at time τ. One can further use the magnitudes of the sensitivity coefficients to rank state variables (at time τ) according to the degree of their influence on a particular state variable (at time t), where larger sensitivity magnitudes indicate higher importance.

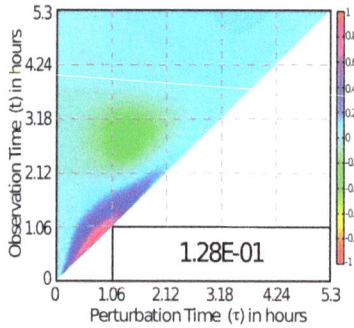

Figure 1. A heatmap of the molecular density function perturbation (MDFP) sensitivity coefficient. The *x*-axis represents the time τ at which the perturbation is introduced while the *y*-axis represents the observation time t. The MDFP coefficient in the heatmap is scaled such that the magnitude falls within ± 1, and the scaling factor is reported in the lower right corner of the plot. The sensitivity values for $t < \tau$ are set to zero for causal systems.

In the case study, we considered an ensemble of ODE models with each model representing one cell in a cell population. The models in the ensemble shared the same ODEs and parameters but had different initial states. The ODE model followed the general formula:

$$\frac{d\mathbf{x}(t, \mathbf{p})}{dt} = g(\mathbf{x}, \mathbf{p}), \tag{8}$$

where $\mathbf{p}$ denotes the vector of model parameters and $g(\mathbf{x}, \mathbf{p})$ is a vector-valued nonlinear function. The distribution of the initial conditions is given by the PDF $f_{\mathbf{X}}(\mathbf{x}_{t_0}, t_0)$. The sensitivity coefficients were computed using a Monte Carlo approach, where we simulated an ensemble of ODE models with a random sample generated from $f_{\mathbf{X}}(\mathbf{x}_{t_0}, t_0)$ as the initial conditions. The model simulation of each randomly sampled initial condition represented the state trajectory of a cell in the cell population. For the computation of $S_{i,j}^{MDFP}(t, \tau)$, we introduced a perturbation δ to the state variable x_j for each of the cells in the ensemble at selected time points τ and simulated the perturbed state trajectory of x_i until the desired time t ($t \geq \tau$). We constructed the marginal PDFs or CDFs of the state variables using a kernel density estimator with leave-one-out cross validation [33].

2.2. Green's Function Matrix Analysis

We compared the MDFP analysis to a related sensitivity analysis based on the GFM. Similar to the MDFP analysis, the GFM analysis introduces time-dependent perturbations to the state variables. The GFM sensitivity coefficients were calculated by directly differentiating the ODE model in Equation (8) as follows [28]:

$$\frac{d}{dt}\left(\frac{\partial \mathbf{x}(t)}{\partial \mathbf{x}(\tau)}\right) = \frac{d}{dt}\mathbf{S}^{GFM}(t, \tau) = \frac{\partial g(\mathbf{x}, \mathbf{p})}{\partial \mathbf{x}}\mathbf{S}^{GFM}; \quad \mathbf{S}^{GFM}(\tau, \tau) = \mathbf{I}_n \tag{9}$$

where $\mathbf{S}^{GFM}(t, \tau)$ is the $n \times n$ sensitivity matrix and $\mathbf{I}_n$ is the $n \times n$ identity matrix. The (i, j)-th element of $\mathbf{S}^{GFM}(t, \tau)$ (i.e., $S_{i,j}^{GFM}(t, \tau) = dx_i(t)/dx_j(\tau)$) gives the sensitivity of the state $x_i(t)$ with respect to perturbations to the state $x_j(\tau)$ ($t \geq \tau$). In the case study below, we normalized the sensitivity coefficients as follows:

$$\hat{S}_{i,j}^{GFM}(t,\tau) = S_{i,j}^{GFM}(t,\tau)\frac{x_j(\tau)}{x_i(t)} \tag{10}$$

We computed the GFM sensitivity coefficients following the procedure described in the original publication [28].

3. Results

3.1. TRAIL-Induced Cell Death Model in HeLa Cells

Figure 2 depicts the signaling network associated with extrinsically-induced apoptosis by the tumor necrosis factor-related apoptosis-inducing ligand (TRAIL). The ODE model comprises 58 species, 28 reactions, and 70 kinetic parameters [34] (see Supplementary Material S2 for details on the initial conditions, parameter values, and rate equations). The model parameters and initial conditions were previously determined by parameter fitting to single-cell and cell population data from cell imaging, flow cytometry, and immunoblotting experiments [31,35,36]. The model describes the key mechanisms for the activation of endogenous executioner caspase-3 (C3*) and the subsequent cleavage of poly(ADP-ribose) polymerase (PARP) [35]. Specifically, the model describes four major pathways: (i) the upstream pathway, describing TRAIL-induced cleavage of pro-caspase-8 (C8) to caspase-8 (C8*); (ii) the mitochondrial independent type I pathway, describing the cleavage of pro-caspase-3 (C3) to caspase-3 (C3*) by C8* and the inhibition of C3 by X-linked inhibitor of apoptosis (XIAP); (iii) the mitochondrial-dependent type II pathway, describing the formation of mitochondrial pores promoted by C8*, the consequent release of cytochrome c (Cyc) into the cytosol, the formation of apoptosome (Apop) induced by cytosolic Cyc, and the activation of C3 by apoptosome; and (iv) the pro-caspase-6 (C6) positive feedback loop where active C3* could promote the activation of C8. In the following, we applied the GFM and MDFP sensitivity analysis to elucidate the key processes in the cell death decision making. More specifically, we computed the GFM and MDFP sensitivity coefficients of the cleaved PARP (cPARP) concentration, an indicator of apoptosis, with respect to perturbations in the concentration of molecules involved in the regulation of PARP cleavage in the model, excluding the intermediate complexes.

3.2. GFM Analysis of TRAIL-Induced Cell Death

We applied the GFM analysis to the ODE model using the model parameters in the original report and the median initial concentration from a follow-up publication by the same authors [36] (see Supplementary Material S2). The analysis was performed for a constant TRAIL stimulation over a time range of 0 to 5.3 h, in which the concentrations reached steady state (see Figure 2b). Here, the ODE model simulated an apoptotic cell in which the cleavage of PARP in response to TRAIL occurs in a delayed switch-like manner, as shown in Figure 3a. To study the activation dynamics of cPARP in greater detail, the analysis of the GFM sensitivity coefficients was split into two phases: before and after mitochondrial outer membrane permeabilization (pre- and post-MOMP). Following a previous study [31], we defined MOMP to occur when 10% of the total PARP has been cleaved into cPARP, which in this analysis occurred at 2.36 h (see Figure 2b). Figure 3b,c portrays the ten largest GFM sensitivity coefficients of the cPARP concentration $\hat{S}_{cPARP,j}^{GFM}(t,\tau)$ in the pre- and post-MOMP phases, respectively (see Supplementary Figures S1 and S2 for the complete GFM sensitivity coefficients). In the pre-MOMP phase, the ten largest GFM sensitivity magnitudes were associated with the upstream and type I pathways, indicating that the early dynamics of cPARP response to the TRAIL stimulus depends on these two pathways. In the post-MOMP phase, the top sensitivity coefficients corresponded to the type II pathway, specifically the regulators of MOMP (i.e., the signaling molecules upstream of M* in the network in Figure 2). Thus, the GFM analysis indicates that the switch-like dynamics in the cPARP concentration relies on the mitochondria-dependent pathway.

(a)

(b)

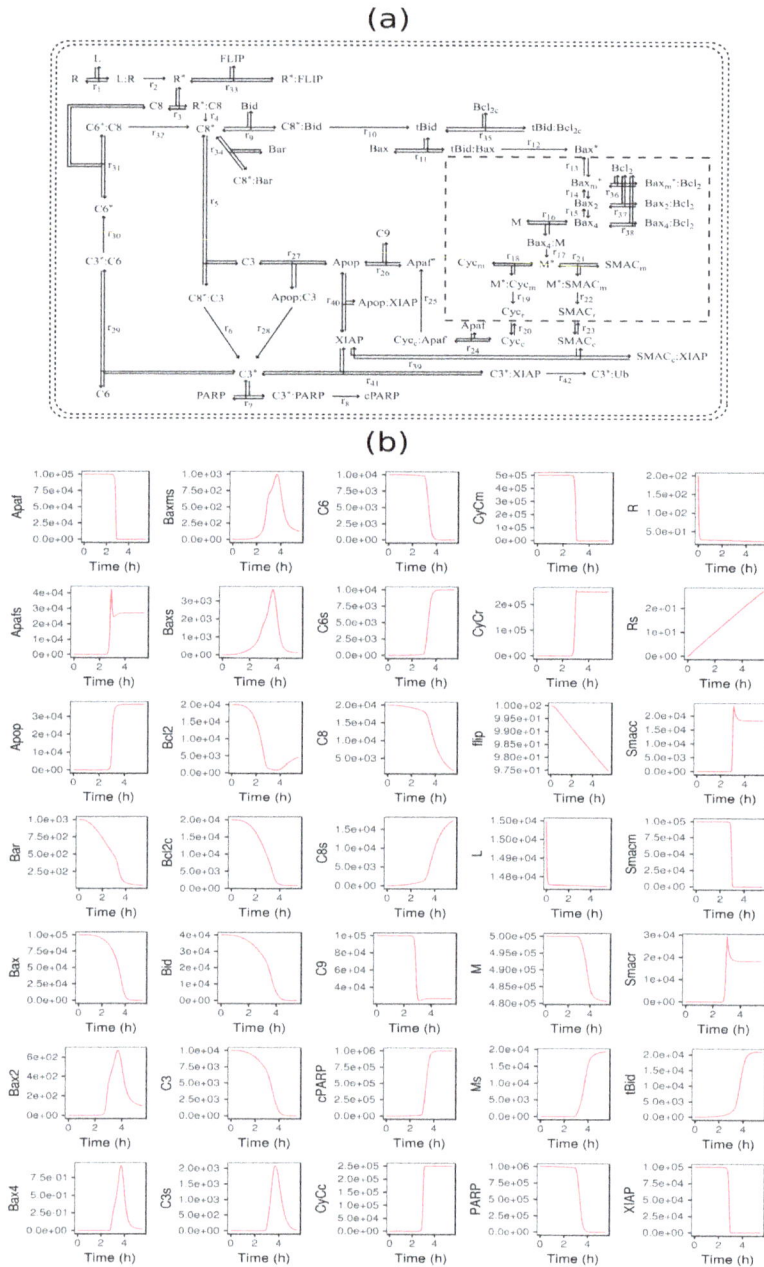

Figure 2. Signal transduction pathway and model simulation of TRAIL (tumor necrosis factor-related apoptosis-inducing ligand)-induced apoptosis in HeLa cells. (**a**) Signal transduction pathway of apoptosis. Type I pathway describes the activation of caspase-3 by caspase-8 while type II pathway describes a mitochondria-dependent activation of caspase-3. Active caspase-3 subsequently cleaves the substrate poly(ADP-ribose) polymerase (PARP) to produce cleaved poly(ADP-ribose) polymerase (cPARP). (**b**) Model simulation of signal transduction pathway in response to TRAIL.

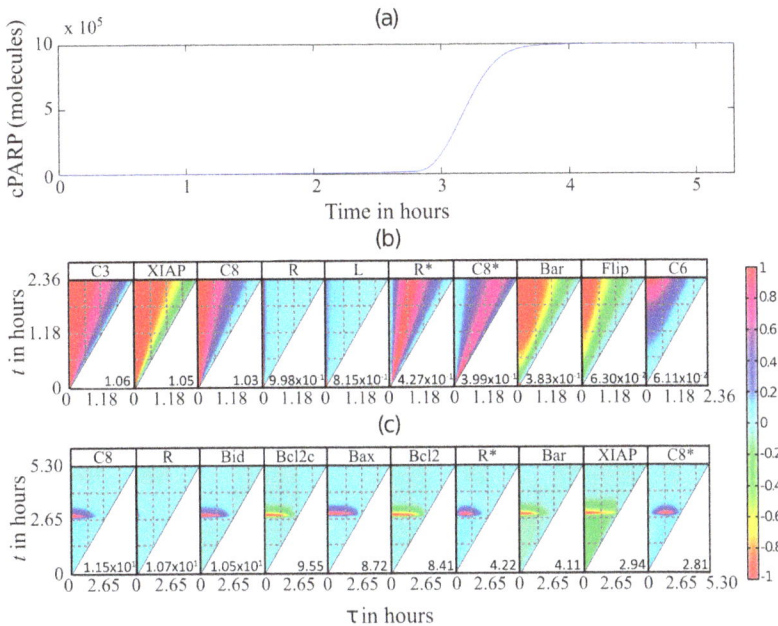

Figure 3. Green's function matrix (GFM) analysis of cPARP activation by a constant TRAIL stimulus. (**a**) cPARP activation follows a delayed switch-like trajectory in response to a constant TRAIL stimulus. (**b**,**c**) Ten largest GFM sensitivity coefficients of cPARP concentration (in magnitude) with respect to perturbations to the state variables in the network, as shown on the label of each subfigure. The *x*-axis gives the time of perturbation τ while the *y*-axis represents the time of observation *t*. Each heatmap is scaled to have values within ±1, using the scaling factor reported in the lower right corner of the plot. Panel (**b**) shows the GFM sensitivity coefficients in the pre-MOMP phase (before 2.36 h). Panel (**c**) shows the GFM sensitivity coefficients in the post-MOMP phase (after 2.36 h).

3.3. MDFP Analysis of TRAIL-Induced Cell Death

The MDFP analysis was carried out for the same TRAIL stimulation as the GFM. For the calculation of the cell distribution, we generated five ensembles of 1000 initial concentrations from a log-normal distribution using the Latin hypercube sampling (LHS) algorithm based on the reported mean values and coefficient of variations reported previously [36] (see Supplementary Material S2). Figure 4a gives the time evolution of the distribution of the cPARP concentration based on the simulations of the ODE model using the ensemble of initial concentrations. Following the original study [36], we defined cells to be apoptotic when 50% of the total PARP at the final time exists in its cleaved form. The ensemble model simulations showed that on average ~95% of the cells in the simulated cell population undergo apoptosis, similar to what was reported in the original modeling study [36].

As in the GFM analysis above, we computed the MDFP sensitivities of cPARP with respect to perturbations to the concentrations of other molecules in the network. Following Equation (2), we introduced a mean shift perturbation to the distribution of each state variable at various perturbation times τ, specifically by adding +10% or -10% of the mean concentration to the state variable $x_j(\tau)$ (i.e., $\delta \mathbf{e_j} = \pm 0.1 \mu_j(\tau) \mathbf{e_j}$, where $\mu_j(\tau)$ is the mean of the state x_j at time τ). Starting from the perturbed concentrations, we simulated the time-evolution of cPARP for time $t \geq \tau$. Based on these simulations, we reconstructed the marginal PDFs and CDFs of the cPARP using the kernel density function approximation, which were then used in the calculation of the sensitivity coefficients as prescribed in Equation (6).

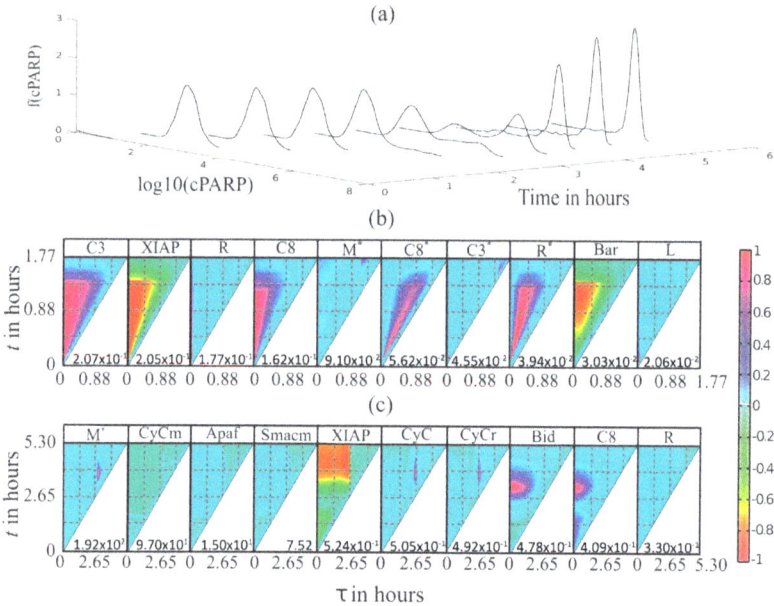

Figure 4. MDFP analysis of cPARP activation by a constant TRAIL stimulus. (**a**) Time evolution of the distribution of cPARP concentration shows a switch-like behavior. (**b,c**) Ten largest MDFP coefficients of cPARP concentration (in magnitude) with respect to the perturbations to different state variables in the network. The *x*-axis gives the time of perturbation τ while the *y*-axis gives the time of observation t. Each heatmap is scaled to have values within ± 1, using the scaling factor reported in the lower right corner of the plot. Panel (**b**) shows the MDFP sensitivity coefficients pre-MOMP (until 1.76 h). Panel (**c**) shows the MDFP sensitivity coefficients post-MOMP.

Figure 4b,c shows the heatmaps of the ten largest MDFP sensitivity coefficients (in magnitude) averaged over the five ensembles (see Supplementary Figures S3 and S4 for the complete MDFP sensitivity coefficients). We also split the analysis into two phases: pre- and post-MOMP at 1.76 h, the time when the median of cPARP concentration reached 10% of the median of total PARP concentration. Similar to the GFM analysis, the MDFP analysis showed that the early response of cPARP to TRAIL-induced apoptosis depends on the upstream and type I pathway molecules. Meanwhile, the cleavage of PARP in the post-MOMP phase is sensitive to mitochondria-dependent pathway molecules, again confirming the general finding of the GFM analysis above. However, in contrast to the GFM analysis, the MDFP sensitivity coefficients pointed to events during and after MOMP, such as the release of cytochrome c from mitochondria, the binding of XIAP by Smac, and the formation of the apoptosome, as the key regulators of cPARP concentration.

3.4. MDFP Analysis of Apoptotic and Non-Apoptotic HeLa Cells

We repeated the MDFP analysis focusing on the subpopulations of apoptotic and non-apoptotic cells separately. Here, the final cPARP concentration (at time 5.3 h) was taken to be the indicator of apoptosis, where an apoptotic cell has at least 50% of the total PARP cleaved (see Figure 5a,b) [36]. Since only 5% of the population was non-apoptotic, a resampling of the initial conditions was performed to simulate 10,000 cells, from which a population of 1000 apoptotic and 1000 non-apoptotic cells were chosen for MDFP analysis. We then ranked the molecules according to the infinite norm of the MDFP sensitivity coefficients of the final cPARP level with respect to the respective molecular concentrations (i.e., $\|S_{cPARP,j}^{MDFP}\|_\infty = \max_\tau S_{cPARP,j}^{MDFP}(5.3\ h, \tau)$). Figure 5c,d shows the ranking of the top ten molecules according to the magnitudes of $\|S_{cPARP,j}^{MDFP}\|_\infty$ (see Supplementary Figure S5 for the complete MDFP sensitivity coefficients of non-apoptotic cells). The MDFP ranking of the apoptotic subpopulation was in agreement with the GFM analysis in which the final cPARP level depended on the molecules that regulate MOMP. The similarity between the GFM and MDFP analyses of an apoptotic subpopulation is perhaps not surprising considering that the GFM analysis was applied to the model of a cell undergoing apoptosis. Meanwhile, the analysis of a non-apoptotic subpopulation produced a ranking that resembled the outcome of the MDFP analysis of the cell population above. Comparing the analysis of the apoptotic and non-apoptotic cells showed the importance of MOMP, XIAP and its inhibitor Smac, and Apaf-1 in regulating the final cPARP in non-apoptotic cells. Interestingly, among the apoptotic cells, XIAP was not among the 10 largest sensitivity coefficients.

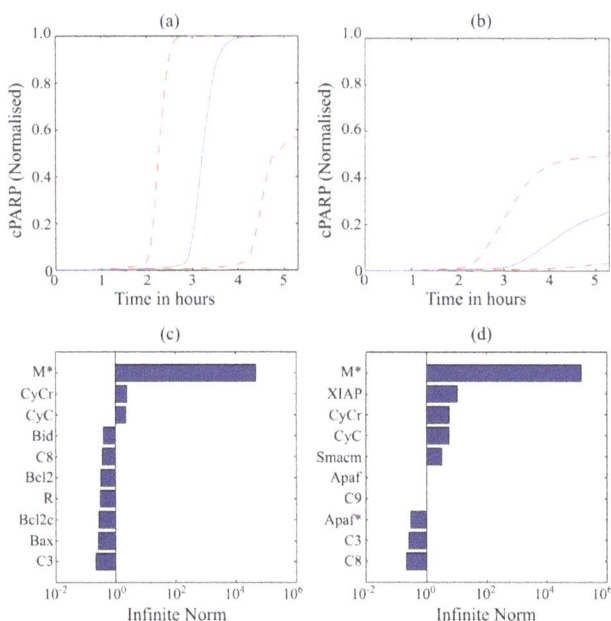

Figure 5. MDFP analysis of the final cleaved PARP levels in (**a**,**c**) apoptotic and (**b**,**d**) non-apoptotic cell subpopulations. (**a**,**b**) The level of cPARP normalized with respect to the total PARP level. The dashed lines (–) indicate the 1 and 99 percentiles of the cPARP levels, while the solid line (-) represents the median level. (**c**,**d**) Ten largest sensitivity coefficients in magnitude in apoptotic and non-apoptotic cells, respectively.

4. Discussion

Cell-to-cell variability has important functional roles in physiological processes, such as cell decision making in stem cell differentiation and cell death. In this work, we developed a sensitivity analysis method called molecular density function perturbation (MDFP) based on introducing time-varying mean shift perturbations to the distribution of molecular concentrations and quantifying the effects of such perturbations on the distribution of the concentration of molecules of interest. The magnitude of the MDFP sensitivity coefficients indicates how much a perturbation to the concentration PDF of one molecule introduced at a particular time τ affects the concentration PDF of a molecule of interest at some time t ($t \geq \tau$). We applied the MDFP analysis to a model of programmed cell death signaling in a population of HeLa cells to elucidate the apoptosis decision making. We used the magnitude of the sensitivity coefficients to rank the importance of each molecule in determining the concentration of cleaved PARP, an indicator of apoptosis.

In the application of the MDFP analysis, we employed the Cramer–von Mises distribution distance Δ_{CVM} in the calculation of the sensitivity coefficients. As mentioned in Section 2.1, several alternative distribution distance metrics exist for defining the MDFP sensitivity coefficients. The rankings of molecules based on the cPARP sensitivity coefficients using different distribution distances were strongly correlated with the Cramer–von Mises and with each other (see Supplementary Figure S6a). Furthermore, the ranking of molecules using different perturbation magnitudes (1%, 10% and 100% of the mean) were in agreement with each other (see Supplementary Figure S6b). Thus, the conclusion of the MDFP analysis did not depend strongly on the choice of distribution distance and perturbation size.

Global sensitivity analysis methods such as SOBOL sensitivity [16], DGSM [17], and glocal analysis [18] can be applied to analyze mathematical models of cell populations. As mentioned earlier and explained in [29], the dynamical aspects of cellular regulation may not be immediately apparent from the application of these analyses. Briefly, the reason stems from the fact that the perturbations in these methods are introduced to model parameters in a time-invariant (static) manner. Consequently, the effects of the perturbations on the system behavior are integrated over time [29]. While existing global sensitivity analyses are able to indicate which parametric perturbations cause a significant change in the overall system behavior (output), the sensitivity coefficients do not directly point to when these perturbations matter.

At the cost of requiring more complicated calculations than the existing global sensitivity analysis methods, the MDFP analysis provides dynamic information on the bottlenecking process by revealing the molecular concentrations to which perturbations introduced at time τ would elicit a large change in a particular state variable of interest at t. For example, referring to Figure 4c, the heatmap of the MDFP sensitivity coefficient of cPARP with respect to pro-caspase-8 (C8) indicates that perturbing the distribution of pro-caspase-8 at the beginning of the experiment $\tau = 0$ (h) would cause a much higher impact on cPARP compared to a perturbation delivered after ~2 h. Figure 6 shows the effects of a positive mean shift perturbation to C8 ($\delta = +\mu_{C8}$) at two different perturbation times, either $\tau = 0$ h or $\tau = 2.14$ h, on the mean, median, and standard deviation of cPARP distribution, confirming the MDFP sensitivity analysis.

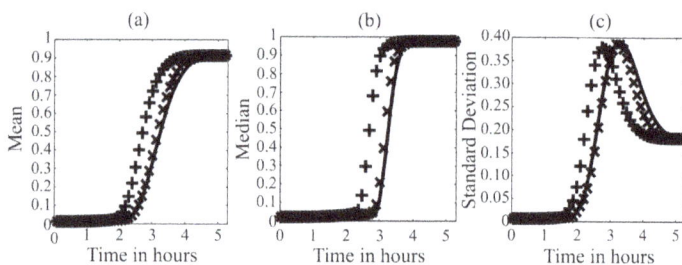

Figure 6. Validation of the MDFP sensitivity analysis of cPARP. A positive mean shift perturbation to pro-caspase-8 was given either at $\tau = 0$ h (+) or at $\tau = 2.14$ h (×). Panel (**a**) shows the mean; panel (**b**) gives the median; and panel (**c**) gives the standard deviation of the cPARP concentration. The unperturbed simulation is shown as solid lines (−).

The MDFP analysis of the cell distribution and the GFM analysis of the ODE model provided somewhat different conclusions with respect to the regulation of PARP cleavage. According to the GFM analysis, the switching dynamics of cPARP depends on the molecules upstream of MOMP, particularly the initial level of pro-caspase-8 (C8). On the other hand, the MDFP analysis suggests that PARP cleavage is strongly sensitive to MOMP and the subsequent release of cytochrome c into the cytosolic compartment. As done in Figure 6, we compared perturbing the mean initial concentration of pro-caspase 8 (C8) at the initial time $\tau = 0$ with perturbing the mean number of mitochondrial open pores (M*) at time $\tau = 2.14$ h, when M* level had reached steady state for more than 99% of the cells. Both perturbations were implemented using 100% positive mean shifts. Figure 7 shows the effects of the above perturbations on the mean, median, and standard deviation of cPARP concentration. As illustrated in Figure 7, both perturbations led to similar shifts in the mean and median of cPARP, where the switch-like dynamic of PARP cleavage occurred earlier and more swiftly. Meanwhile, the perturbation to M* caused a larger drop in the standard deviation of cPARP than the perturbation to C8 (i.e., cells became more alike when we increased the number of mitochondrial open pores). While the positive mean shift perturbation to pro-caspase-8 led to a faster cleavage of PARP, this perturbation did not affect the fraction of apoptotic versus non-apoptotic cells. However, when we increased the number of mitochondrial open pores, the fraction of non-apoptotic cells dropped from 5.6% to 3.2%.

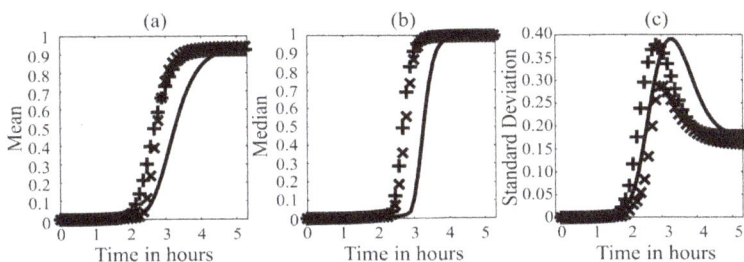

Figure 7. Comparison of GFM and MDFP analyses. A positive mean shift perturbation was given either to pro-caspase-8 a $\tau = 0$ h (+) or to mitochondrial open pores M* at $\tau = 2.14$ h ($\times$). Panel (**a**) shows the mean of the cPARP concentration distribution, panel (**b**) gives the median, and panel (**c**) gives the standard deviation. The unperturbed simulation is shown as solid lines ($-$).

Both the GFM and MDFP analyses of the cell death signaling network implicate the mitochondria-dependent type II pathway to be the responsible mechanism in the switch-like activation of PARP in HeLa cells, placing caspase-8 activation (cleavage) as the most important step in the apoptosis decision making during pre-MOMP. This finding agrees with a previous experimental study on fractional killing by TRAIL [31,37], reporting that the activation of C8 controls the switching time of cPARP. In post-MOMP, the GFM analysis indicates that perturbations to the regulators of MOMP (upstream of M* in Figure 2) would strongly affect the PARP cleavage dynamics. On the other hand, the MDFP analysis points to MOMP and events post-MOMP (downstream of M* in Figure 2), including cytochrome c and Smac release from mitochondria, XIAP binding by Smac, and apoptosome formation, to be the key determinants of the cell-to-cell variability in cPARP level. The finding from the MDFP analysis is in agreement with a previous study that found XIAP to be the determining factor for the rate and extent of type II cell death [38]. Furthermore, the results of the MDFP analysis of apoptotic and non-apoptotic subpopulations showed that perturbations to molecules executing the cell death signal after MOMP (i.e., Smac, cytochrome C, Apaf-1) had a stronger effect on the cPARP activation in the non-apoptotic cells than in the apoptotic cells. Consistent with such an insight, the depletion of Apaf-1 or Apaf-1/Smac together by siRNA has been shown to significantly reduce the activation of PARP in HeLa cells (see Supplementary Figure S7 in [36]).

As the functional significance of cell-to-cell variability is increasingly being recognized and the mathematical models that are able to describe cell distribution become more and more common, the MDFP analysis proposed here will provide an analytical tool to use such models for elucidating the key molecules and processes that govern the dynamics of cellular heterogeneity.

Supplementary Materials: The following are available online at www.mdpi.com/2227-9717/6/2/9/s1, Material S1: Probability distance metrics, Material S2: TRAIL induced programmed cell death model of Hela cells, Figure S1: GFM analysis of TRAIL induced apoptosis model during pre-MOMP (before 2.36 h), Figure S2: GFM analysis of TRAIL-induced apoptosis model during post-MOMP (after 2.36 h), Figure S3: MDFP analysis of TRAIL-induced apoptosis model during pre-MOMP (before 1.76 h), Figure S4: MDFP analysis of TRAIL-induced apoptosis model during post-MOMP (after 1.76 h), Figure S5: MDFP analysis of non-apoptotic Hela subpopulation, Figure S6: Spearman correlations of MDFP sensitivity coefficients using different distribution distances and perturbation sizes.

Acknowledgments: TMP was supported by the Singapore Millennium Foundation scholarship. The authors would like to acknowledge funding from ETH Zurich, as well as support from Sage Bionetworks.

Author Contributions: T.M.P. and R.G. conceived and designed the study; T.M.P. performed the computational work; T.M.P. and R.G. analyzed the data; T.M.P. and R.G. wrote the paper.

Conflicts of Interest: The authors declare no conflict of interest. The founding sponsors had no role in the design of the study; in the collection, analyses, or interpretation of data; in the writing of the manuscript, or in the decision to publish the results.

References

1. Cahan, P.; Daley, G.Q. Origins and implications of pluripotent stem cell variability and heterogeneity. *Nat. Rev. Mol. Cell Biol.* **2013**, *14*, 357–368. [CrossRef] [PubMed]
2. Flusberg, D.A.; Sorger, P.K. Surviving apoptosis: Life-death signaling in single cells. *Trends Cell Biol.* **2015**, *25*, 446–458. [CrossRef] [PubMed]
3. Xia, X.; Owen, M.S.; Lee, R.E.C.; Gaudet, S. Cell-to-cell variability in cell death: Can systems biology help us make sense of it all? *Cell Death Dis.* **2014**, *5*, e1261. [CrossRef] [PubMed]
4. Golding, I.; Paulsson, J.; Zawilski, S.M.; Cox, E.C. Real-time kinetics of gene activity in individual bacteria. *Cell* **2005**, *123*, 1025–1036. [CrossRef] [PubMed]
5. Raj, A.; Peskin, C.S.; Tranchina, D.; Vargas, D.Y.; Tyagi, S. Stochastic mRNA synthesis in mammalian cells. *PLoS Biol.* **2006**, *4*, e309. [CrossRef] [PubMed]
6. Raj, A.; van Oudenaarden, A. Nature, nurture, or chance: Stochastic gene expression and its consequences. *Cell* **2008**, *135*, 216–226. [CrossRef] [PubMed]
7. Jia, G.; Stephanopoulos, G.; Gunawan, R. Ensemble kinetic modeling of kinetic metabolic networks from dynamics metabolic profiles. *Metabolites* **2012**, *2*, 891–912. [CrossRef] [PubMed]
8. Stamakis, M. Cell population balance and hybrid modeling of population dynamics for a single gene with feedback. *Comput. Chem. Eng.* **2013**, *53*, 25–34. [CrossRef]
9. Hasenauer, J.; Hasenauer, C.; Hucho, T.; Theis, F.J. ODE constrained mixture modelling: A method for unraveling subpopulation structures and dynamics. *PLoS Comput. Biol.* **2014**, *10*, e1003686. [CrossRef] [PubMed]
10. Henson, M.A. Dynamic modeling of microbial cell populations. *Curr. Opin. Biotechnol.* **2003**, *14*, 460–467. [CrossRef]
11. Hasty, J.; Pradines, J.; Dolnik, M.; Collins, J.J. Noise-based switches and amplifiers for gene expression. *Proc. Natl. Acad. Sci. USA* **2000**, *97*, 2075–2080. [CrossRef] [PubMed]
12. Manninen, T.; Linne, M.L.; Ruohonen, K. Developing Ito stochastic differential equation models for neuronal signal transduction pathways. *Comput. Biol. Chem.* **2006**, *30*, 280–291.
13. Samoilov, M.S.; Arkin, A.P. Deviant effects in molecular reaction pathways. *Nat. Biotechnol.* **2006**, *24*, 1235–1240. [CrossRef] [PubMed]
14. Shahrezaei, V.; Swain, P.S. Analytical distributions for stochastic gene expression. *Proc. Natl. Acad. Sci. USA* **2008**, *105*, 17256–17261. [CrossRef] [PubMed]
15. Poovathingal, S.K.; Gruber, J.; Halliwell, B.; Gunawan, R. Stochastic drift in mitochondrial DNA point mutations: A novel perspective *ex silico*. *PLoS Comput. Biol.* **2009**, *5*, e1000572. [CrossRef] [PubMed]
16. Saltelli, A.; Ratto, M.; Andres, T.; Campolongo, F.; Cariboni, J.; Gatelli, D.; Saisana, M.; Tarantola, S. *Global Sensitivity Analysis. The Primer*; John Wiley & Sons: Hoboken, NJ, USA, 2008.

17. Kucherenko, S.; Rodriguez-Fernandez, M.; Pantelides, C.; Shah, N. Monte Carlo evaluation of derivative-based global sensitivity measures. *Reliab. Eng. Syst. Saf.* **2009**, *94*, 1135–1148. [CrossRef]

18. Hafner, M.; Koeppl, H.; Hasler, M.; Wagner, A. "Glocal" Robustness Analysis and Model Discrimination for Circadian Oscillators. *PLoS Comput. Biol.* **2009**, *5*, e1000534. [CrossRef] [PubMed]

19. Marino, S.; Hogue, I.B.; Ray, C.J.; Kirschner, D.E. A methodology for performing global uncertainty and sensitivity analysis in systems biology. *J. Theor. Biol.* **2008**, *254*, 178–196. [CrossRef] [PubMed]

20. Gunawan, R.; Cao, Y.; Petzold, L.; Doyle, F.J., III. Sensitivity analysis of discrete stochastic systems. *Biophys. J.* **2005**, *88*, 2530–2540. [CrossRef] [PubMed]

21. Komorowski, M.; Costa, M.J.; Rand, D.A.; Stumpf, M.P.H. Sensitivity, robustness, and identifiability in stochastic chemical kinetics models. *Proc. Natl. Acad. Sci. USA* **2011**, *108*, 8645–8650. [CrossRef] [PubMed]

22. Plyasunov, S.; Arkin, A.P. Efficient stochastic sensitivity analysis of discrete event systems. *J. Comput. Phys.* **2007**, *221*, 724–738. [CrossRef]

23. Rathinam, M.; Sheppard, P.W.; Khammash, M. Efficient computation of parameter sensitivities of discrete stochastic chemical reaction networks. *J. Chem. Phys.* **2010**, *132*, 034103. [CrossRef] [PubMed]

24. Gunawan, R.; Jung, M.Y.L.; Braatz, R.D.; Seebauer, E.G. Parameter sensitivity analysis applied to modeling of transient enhanced diffusion and activation of Boron in Silicon. *J. Electrochem. Soc.* **2003**, *150*, G758–G765. [CrossRef]

25. Gunawan, R.; Doyle, F.J., III. Phase sensitivity analysis of circadian rhythm entrainment. *J. Biol. Rhythms* **2007**, *22*, 180–194. [CrossRef] [PubMed]

26. Ingalls, B. Sensitivity analysis: From model parameters to system behaviour. *Essays Biochem.* **2008**, *45*, 177–193. [CrossRef] [PubMed]

27. Zi, Z. Sensitivity analysis approaches applied to systems biology models. *IET Syst. Biol.* **2011**, *5*, 336–346. [CrossRef] [PubMed]

28. Perumal, T.M.; Wu, Y.; Gunawan, R. Dynamical analysis of cellular networks based on the Green's function matrix. *J. Theor. Biol.* **2009**, *261*, 248–259. [CrossRef] [PubMed]

29. Perumal, T.M.; Gunawan, R. Understanding dynamics using sensitivity analysis: Caveat and solution. *BMC Syst. Biol.* **2011**, *5*, 41. [CrossRef] [PubMed]

30. Perumal, T.M.; Krishna, S.M.; Tallam, S.S.; Gunawan, R. Reduction of kinetic models using dynamic sensitivities. *Comput. Chem. Eng.* **2013**, *56*, 37–45. [CrossRef]

31. Spencer, S.L.; Gaudet, S.; Albeck, J.G.; Burke, J.M.; Sorger, P.K. Non-genetic origins of cell-to-cell variability in TRAIL-induced apoptosis. *Nature* **2009**, *459*, 428–432. [CrossRef] [PubMed]

32. Varma, A.; Morbidelli, M.; Wu, H. *Parametric Sensitivity in Chemical Systems*; Cambridge University Press: Cambridge, UK, 1999.

33. Peter, D.H. Kernel estimation of a distribution function. *Commun. Stat. Theory Methods* **1985**, *14*, 605–620. [CrossRef]

34. Niepel, M.; Spencer, S.L.; Sorger, P.K. Non-genetic cell-to-cell variability and the consequences for pharmacology. *Curr. Opin. Chem. Biol.* **2009**, *13*, 556–561. [CrossRef] [PubMed]

35. Albeck, J.G.; Burke, J.M.; Aldridge, B.B.; Zhang, M.; Lauffenburger, D.A.; Sorger, P.K. Quantitative analysis of pathways controlling extrinsic apoptosis in single cells. *Mol. Cell* **2008**, *30*, 11–25. [CrossRef] [PubMed]

36. Albeck, J.G.; Burke, J.M.; Spencer, S.L.; Lauffenburger, D.A.; Sorger, P.K. Modeling a snap-action, variable-delay switch controlling extrinsic cell death. *PLoS Biol.* **2008**, *6*, 2831–2852. [CrossRef] [PubMed]

37. Roux, J.; Hafner, M.; Bandara, S.; Sims, J.J.; Hudson, H.; Chai, D.; Sorger, P.K. Fractional killing arises from cell-to-cell variability in overcoming a caspase activity threshold. *Mol. Syst. Biol.* **2015**, *11*, 803. [CrossRef] [PubMed]

38. Gaudet, S.; Spencer, S.L.; Chen, W.W.; Sorger, P.K. Exploring the contextual sensitivity of factors that determine cell-to-cell variability in receptor-mediated apoptosis. *PLoS Comput. Biol.* **2012**, *8*, e1002482. [CrossRef] [PubMed]

Article

Genome-Scale *In Silico* Analysis for Enhanced Production of Succinic Acid in *Zymomonas mobilis*

Hanifah Widiastuti [1,†], Na-Rae Lee [2,†], Iftekhar A. Karimi [1] and Dong-Yup Lee [1,2,3,4,*]

[1] Department of Chemical and Biomolecular Engineering, National University of Singapore, 4 Engineering Drive 4, Singapore 117585, Singapore; dongyuplee@gmail.com (H.W.); cheiak@nus.edu.sg (I.A.K.)

[2] NUS Synthetic Biology for Clinical and Technological Innovation (SynCTI), Life Sciences Institute, National University of Singapore, 28 Medical Drive, Singapore 117456, Singapore; bchlnr@nus.edu.sg

[3] Bioprocessing Technology Institute, Agency for Science, Technology and Research (A*STAR), 20 Biopolis Way, #06-01, Centros, Singapore 138668, Singapore

[4] School of Chemical Engineering, Sungkyunkwan University, 2066 Seobu-ro, Jangan-gu, Suwon, Gyeonggi-do 16419, Korea

* Correspondence: dongyuplee@skku.edu; Tel.: +82-31-290-7253
† These authors contributed equally to this work.

Received: 18 January 2018; Accepted: 26 March 2018; Published: 1 April 2018

Abstract: Presented herein is a model-driven strategy for characterizing the production capability of expression host and subsequently identifying targets for strain improvement by resorting to network structural comparison with reference strain and *in silico* analysis of genome-scale metabolic model. The applicability of the strategy was demonstrated by exploring the capability of *Zymomonas mobilis*, as a succinic acid producer. Initially, the central metabolism of *Z. mobilis* was compared with reference producer, *Mannheimia succiniciproducens*, in order to identify gene deletion targets. It was followed by combinatorial gene deletion analysis. Remarkably, resultant *in silico* strains suggested that knocking out *pdc*, *ldh*, and *pfl* genes encoding pyruvate-consuming reactions as well as the *cl* gene leads to fifteen-fold increase in succinic acid molar yield. The current exploratory work could be a promising support to wet experiments by providing guidance for metabolic engineering strategies and lowering the number of trials and errors.

Keywords: *Zymomonas mobilis*; succinic acid; gene deletion; genome-scale metabolic model; systems biology

1. Introduction

Succinic acid (SA) is a valuable specialty chemical with a wide range of applications in food, pharmaceutical, and chemical industries, currently being synthesized through petrochemical processes by the catalytic hydrogenation of petroleum-derived maleic anhydride [1]. However, the increase in oil prices along with raising environmental awareness has made bio-based SA production an attractive option. Thus, a multitude of studies have attempted to develop SA-producing microbial fermentation systems coupled with the use of renewable biomasses [2,3]. Several well-known SA producers include *Anaerobiospirillum succiniciproducens*, *Actinobacillus succinogenes*, *Mannheimia succiniciproducens*, and recombinant *Escherichia coli*. Generally, SA is excreted during anaerobic fermentation, which has some operational disadvantages such as low cell growth and slow carbon uptake, leading to its low productivity. In addition, these producers secrete other competing fermentative byproducts, e.g., acetic acid, lactic acid, and formic acid, thus reducing SA yield and making its purification difficult.

To overcome the above limitations, researchers have tried to develop various metabolic engineering strategies to enhance SA production mostly in *E. coli* [4–8]. For example, yield was significantly increased by amplifying the enzymatic reactions involved in SA pathway [9–11] and by introducing non-indigenous enzymes to *E. coli* [12,13]. Other interesting options are the inactivation of pathways that compete with SA [14,15] and the construction of glyoxylate cycle to enable aerobic succinate production [16–18]. For other producers, several studies were also carried out to reduce byproduct formations in *A. succiniciproducens* [19,20], *A. succinogenes* [21–23], and *M. succiniciproducens* [24,25].

In summary, various strategies described above have successfully enhanced SA yields and reduced byproduct formations. Nonetheless, there still remain issues regarding slow carbon uptake under anaerobic condition leading to low productivity. Therefore, we can consider other microorganisms that can grow well under anaerobic condition. One potential candidate is *Zymomonas mobilis* that is known to have relatively high catabolic and high glucose uptake rates, producing ethanol as the major fermentative product [26]. Moreover, Swings and DeLey [27] observed that SA was secreted in *Z. mobilis* as a minor byproduct at the yield of 0.0025–0.014 (g/g), when yeast extract was added into the medium. This evidence in conjunction with the operational advantages of anaerobic fermentation motivated us to explore metabolic engineering strategies for substantially enhancing SA production.

One strategy for improving strain performance is to identify potential targets for genetic modification. To this end, gene (or reaction) targets can be conventionally identified by random mutagenesis, followed by intelligent screening. Recently, various omics data generated by high-throughput experimental techniques have been exploited for gene identification through comparative genomic, transcriptomic, proteomic, and metabolomic studies within the context of systems biotechnology [28–32]. Apart from this high-throughput omics profiling, *in silico* analysis of genome-scale metabolic models of various microbes have been successfully employed to investigate their cellular behavior, pose new engineering hypotheses, and identify and evaluate potential genetic targets for strain improvements [33–39]. Several *in silico* analysis methods for strain optimization have been designed to select such candidates for genetic manipulations (addition and/or deletion), thereby giving rise to enhanced productivity and cellular properties. They include OptKnock [40], OptStrain [41], OptGene [42], OptForce [43], Flux-sum analysis [44], and cofactor modification analysis [45,46]. These studies have shown that genome-scale models can assist experimental studies for strain improvement [47]. In this work, we employ our previously developed genome-scale model for *Z. mobilis* and perform *in silico* analyses to explore and evaluate various strain engineering strategies to enhance SA production in *Z. mobilis*. In particular, we use constraint-based flux analyses for gene knockout simulations to develop optimization strategies for overproducing SA.

2. Materials and Methods

2.1. Genome-Scale Metabolic Model of Z. mobilis

The genome-scale metabolic model for *Z. mobilis* ZM4 (ATCC31821), *i*ZM411, has been developed on the basis of its genome annotations [48]. This previous model was further improved using the updated information from KEGG and BioCyc, resulting in 738 reactions and 705 metabolites in 49 specific pathways or subsystems based on their functional roles. It includes major metabolic pathways, i.e., central metabolism, amino acid biosynthesis, lipid metabolism, cell wall metabolism, and vitamin and cofactor metabolism along with the necessary transport reactions for extracellular metabolites. Biomass equation is also derived from the drain of various biosynthetic precursors and relevant cofactors into *Z. mobilis* biomass at their appropriate ratios to quantify the cell specific growth rate. The complete updated model and biomass equation could be found in Supplementary 1 and 2, respectively.

2.2. Constraints-Based Flux Analysis

Mass balance on various metabolites under the steady state assumption gives rise to a set of linear equations with reactions fluxes as unknown quantities. It is mathematically constructed as an underdetermined system since the number of metabolites constraints is less than the number of unknown fluxes to be determined. Thus, unknown fluxes can then be evaluated by applying linear programming to find optimum value for a given cellular objective, subjected to known uptake rates as the constraints in the model. In this study, the objective function considered was maximization of biomass growth to describe the physiological behavior under the growing condition. Various *in silico* analyses were carried out by using MetaFluxNet [49] and CPLEX 12.1.0 solver under the general algebraic modeling system (GAMS) [50].

2.3. Combinatorial Knockout Simulation

Combinatorial knockout analysis was performed by applying additional constraints to the current linear flux model. We removed individual reaction(s) associated with the gene(s) being knocked out from *in silico* models, and evaluated the alteration in flux distribution as the consequence of the deletion of reaction(s). For this analysis, anaerobic growth on glucose was assumed by setting glucose and oxygen uptake rates at 10 mmol/gDCW/h and zero, respectively. Biomass maximization for a steady-state metabolic network, comprised of a set $N = \{1, \ldots, n\}$ of metabolites and a set $M = \{1, \ldots, m\}$ of reactions, can be expressed in the following linear programming (LP) problem:

$$\text{maximize } v_{biomass}$$

$$\text{subject to } \sum_{j \in M} S_{ij} v_j = 0, \ \forall i \in N$$

$$(1 - y_j) lb_j \leq v_j \leq (1 - y_j) ub_j, y_j = \{0, 1\}, \ \forall j \in K, \ K = \text{set of KO candidates}$$

$$\sum_{j \in K} y_j = k, \ \forall j \in K, \ k = \text{number of genes deleted}$$

where S_{ij} is the stoichiometric coefficient of metabolite i in reaction j, v_j is flux value reaction j, and y_j is binary variable, with 1 for deleted reaction and 0 for active reaction. Gene deletion conducted in this study was for single ($k = 1$), double ($k = 2$), triple ($k = 3$), and quadruple ($k = 4$).

2.4. Flux-Sum Quantification

In order to quantify turnover rate of intermediate metabolites, flux-sum (Φ) can be defined as the summation of incoming or outgoing fluxes of given metabolite i as follows [44]:

$$\Phi_i = \sum_{j \in P_i} S_{ij} v_j = -\sum_{j \in C_i} S_{ij} v_j = \frac{1}{2} \sum_j |S_{ij} v_j|$$

where P_i and C_i denote the sets of reactions producing and consuming metabolite i, respectively. Under the pseudo-steady state assumption, i is the mass flow contributed by all of the fluxes producing (consuming) metabolite i.

3. Results and Discussion

The current study aimed at exploring *Z. mobilis* as potential SA producer or expression host organism and developing the *in silico* model-driven systematic strategy for strain improvement. To this end, we first evaluated the potential of *Z. mobilis* to produce SA by computing its theoretical yield using our genome-scale model. Then, we compared and contrasted the central metabolic networks of *Z. mobilis* and a known SA producer, *M. succiniciproducens*, to identify engineering targets for enhancing SA production. This network comparison allowed us to find a few target genes/reactions to be added/deleted, which were then evaluated via *in silico* analysis of the model as successfully demonstrated by Lee et al. [51]. Additionally, having narrowed down the pool of possible targets from the central metabolic reactions, combinatorial knockout analyses were carried out to obtain a list of possible candidates for genetic modification and to investigate their KO effects on SA production. We now describe and discuss each of the above steps in detail.

3.1. Exploring Metabolic Capabilities for Succinic Acid Production in Z. mobilis

As mentioned earlier, Swings and Deley [27] experimentally observed SA yield of 0.0038–0.024 mol per mol glucose in wild-type *Z. mobilis*. Recently, Seo et al. [52] have shown the existence of *frdABCD* genes associated with succinic acid dehydrogenase which is responsible for converting fumarate to SA, from the genome sequence of *Z. mobilis*. This clearly suggests the fumarate reduction to SA is a plausible biosynthetic pathway in *Z. mobilis*. This route includes malic enzyme that converts pyruvate (PYR) to malate (MAL), which is then followed by fumarate hydratase and SA dehydrogenase. This is indeed the pathway present in *M. succiniciproducens* which is well-known succinic acid producer. Thus, our genome-scale model includes both pathways for SA production, and interestingly subsequent simulations of wild-type *Z. mobilis* on glucose confirmed that it naturally secreted SA at 0.06 molar yield during its exponential growth phase, as the byproducts in lysine and methionine biosynthesis. This yield is comparable to the yield observed by Swings and Deley [27].

Next, we maximized SA production in our *in silico* model satisfying the required cellular growth under growing condition in the wild-type *Z. mobilis*. Simulation results showed that *Z. mobilis* was capable of theoretically producing as much as 1.6 mol SA per mol glucose. Furthermore, an increase in SA production led to a decrease in ethanol production, which clearly suggests that ethanol is a competing metabolite for SA production. To understand this SA production further, we explored how the fluxes are diverted from ethanol to SA in this scenario of maximum SA yield. Comparing the flux distribution in this scenario with the growing condition of the wild-type showed the increase in fluxes through tricarboxylic acid (TCA) cycle and carbon dioxide (CO_2) uptake, while the decrease in fluxes through ethanol production pathway was observed. The attenuation of these latter fluxes is obvious again, as ethanol is a competing metabolite for SA. Interestingly, the increased uptake of CO_2 points to its consumption in SA production. This can be inferred from the increased flux of CO_2-consuming malic enzyme that functions as the bridging reaction between glycolysis and TCA cycle in *Z. mobilis*.

3.2. Central Metabolism Comparison with SA Producer to Identify Gene Candidates in Z. mobilis

Of the several known SA producers, *M. succiniciproducens* has the highest SA yield of 1.2 mol per mol glucose, which is 60% of its theoretical SA yield [24]. Both *Z. mobilis* and *M. succiniciproducens* are anaerobic fermentative organisms without the phospho-transferring systems (PTS). Instead, they both utilize transportation protein for exchanging substrates between the cytoplasm and the surrounding environment via a process known as facilitated diffusion [14,53]. These common characteristics suggest that SA production in *Z. mobilis* can be enhanced by imitating *M. succiniciproducens*. Therefore, we compared the central metabolic pathways of *Z. mobilis* and *M. succiniciproducens* to identify gene candidates that may possibly impact fermentation profiles. One significant difference in their metabolisms is that *Z. mobilis* utilizes the Entner–Doudoroff (ED) pathway, while *M. succiniciproducens* has the Embden–Meyerhof–Parnas (EMP) pathway. Since the ED pathway in *Z. mobilis* is one major factor for its high catabolic rate, we focused more on the reactions that are uniquely present in *Z. mobilis*, but absent in *M. succiniciproducens*, which constitute potential gene deletion targets.

As illustrated in Figure 1, pyruvate decarboxylase (*pdc*, PYR → ACALD + CO_2), acetolactate synthase (*als*, PYR → ALAC + CO_2), and citrate lyase (*cl*, CIT → AC + OAA) are three genes that are unique to *Z. mobilis*. Therefore, we created *in silico* mutants without *pdc*, *als*, and *cl* to mimic *M. succiniciproducens* metabolism. The *in silico* simulation results of the wild-type strain and the knockout mutants are shown in Table 1. The deletion of *pdc* increased SA production to 0.5 mol SA per mol glucose. Not surprisingly, *pdc* is crucial for ethanol fermentation in *Z. mobilis*. Our simulations showed that its deletion diverts the flux from ethanol to acetic acid and SA in the TCA cycle on one hand, and the production of lactic acid and formic acid on the other. However, the deletion of acetolactate synthase (*als*) resulted in zero growth, because this gene is essential for valine synthesis in *Z. mobilis*. Clearly, this is not a viable candidate for strain improvement.

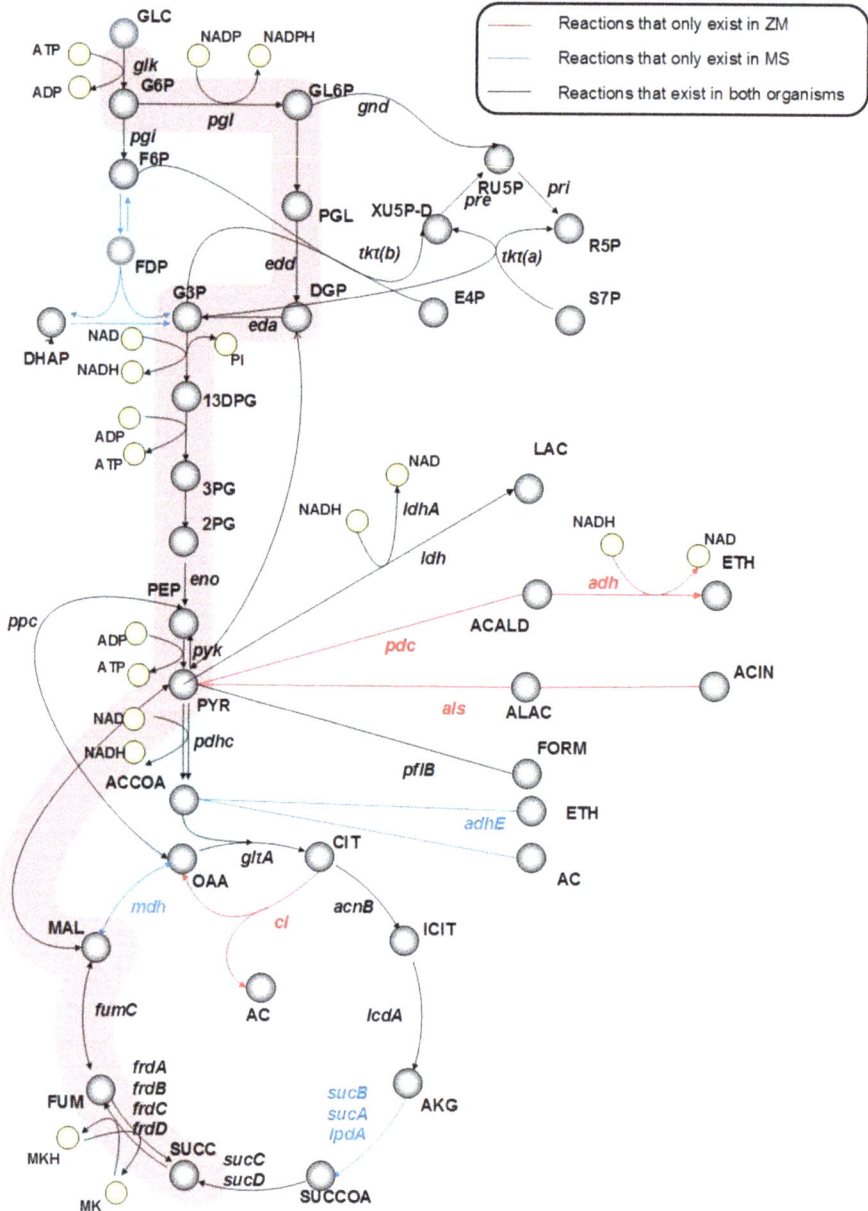

Figure 1. Comparison of central metabolism in *Zymomonas mobilis* and *Mannheimia succiniciproducens*. Blue and red shadows highlight the succinic acid production pathway in *M. succinicproducens* and *Z. mobilis*, respectively.

The deletion of citrate lyase (*cl*) did not affect SA production. But, interestingly, the deletion of *cl* in addition to that of *pdc* increased SA production further from 0.5 to 0.6 mol SA per mol glucose. While this yield is 1.1 mol lower than the maximum possible yield predicted by our *in silico* model for *Z. mobilis*, it is much higher than 0.06 mol observed for the wild-type *Z. mobilis*. This clearly shows the success of a simple approach of mimicking *M. succiniciproducens* for enhancing SA production in *Z. mobilis* as similarly reported in the previous study [9]. However, the question is whether it is possible to increase this yield even more. Therefore, we now use a combinatorial approach in the following section to identify better knockout combinations.

Table 1. Knockout simulations for target genes identified from central metabolism comparison.

In Silico Strains	Growth Rate (1/h)	Molar Yield (mol Metabolite/mol Glucose)				
		Succinic Acid	Ethanol	Lactic Acid	Formic Acid	Acetic Acid
Wild-Type ZM4	0.08	0.06	1.76	0	0	0
Metabolism Comparison						
Δ*pdc*	0.07	0.53	0	0.49	0.28	0.42
Δ*als*	0	0	0	0	0	0
Δ*cl*	0.08	0.06	1.76	0	0	0
Δ*pdc*Δ*cl*	0.07	0.6	0	0.71	0.28	0
Combinatorial Knockout						
Δ*pdc*	0.07	0.53	0	0.49	0.28	0.42
Δ*pdc*Δ*ldh*	0.07	0.93	0	0	1.11	0.56
Δ*pdc*Δ*pfl*	0.07	0.67	0	1.12	0	0
Δ*pdc*Δ*cl*	0.07	0.6	0	0.71	0.28	0
Δ*pdc*Δ*ldh*Δ*pfl*	0.07	1.14	0	0	0	0.75
Δ*pdc*Δ*pfl*Δ*cl*	0.07	0.47	0	1.42	0	0
Δ*pdc*Δ*ldh*Δ*cl*	0.07	1.09	0	0	1.64	0
Δ*pdc*Δ*ldh*Δ*pfl*Δ*cl*	0.07	1.52	0	0	0	0
OptKnock/OptGene						
Δ*pdc*	0.07/0.07	0.53/0.66	0/0	0.59/0.44	0.59/0.94	0
Δ*pdc*Δ*ldh*	0.07/0.07	0.94/0.92	0/0	0.42/0.45	0/0	0.38/0.44
Δ*pdc*Δ*ldh*Δ*pfl*	0.07/0.07	1.17/1.17	0/0	0/0	0/0	0.58/0.45
Δ*pdc*Δ*ldh*Δ*pfl*Δ*cl*	0.07/0.07	1.42/1.42	0/0	0/0	0/0	0

3.3. Combinatorial Knockout Analysis

It should be noted that SA is a primary metabolite of *Z. mobilis*. Therefore, our possible knockout candidates are limited to the 44 genes for the 57 reactions within the central metabolism. Out of the 44 genes, 19 are lethal genes; hence only 25 genes were considered for strain improvement. Since this number is small, we studied all possible combinations of single, double, triple, and quadruple gene knockouts. The resulting predictions are presented in Figure 2.

From single gene knockouts, we find that *pdc* is the only gene whose knockout leads to the increase in SA production at the expense of growth. Without knocking out *pdc*, it is not possible to increase SA production. However, after *pdc* is deleted, several other knockout strategies can be considered to increase SA production further. Not expectedly, cell growth is not affected at all by these additional knockouts, which is a very useful characteristic of *Z. mobilis*. From Figure 2, we see that the additional deletions of *ldh* (lactate dehydrogenase), *pfl* (pyruvate formate lyase), and *cl* all increase SA production. Of these three 2-gene knockouts, *pdc*-*ldh* seems to be the best for SA production with a molar yield of 0.9. Note that Lee et al. [54] had also reported *pdc* and *ldh* as KO candidates for SA production and Seo et al. [55] improved SA production by removing the genes in *Z. mobilis*. We clarified that this is only the best 2-gene knockout with a yield of 0.9 mol per mol of glucose. Further improvement is possibly achievable through 3-gene knockouts. As seen from Figure 2, *pdc*–*ldh*–*pfl* seems to be the best 3-gene knockout with an SA molar yield of 1.1 mol per mol glucose. It should be noted that disrupting *ldh* and *pfl* genes led to the enhanced SA production in *M. succiniciproducens* and *E. coli* [24,56]. Finally, the knockout of all four genes (*pdc–ldh–pfl–cl*) enables us to obtain an even higher yield of 1.5 mol SA per mol glucose, which is very close to the maximum theoretical yield of 1.6 mol SA per mol glucose. Interestingly, the yield of SA was significantly increased by knockout of additional target, i.e., *cl* gene in *Aspergillus niger* [57].

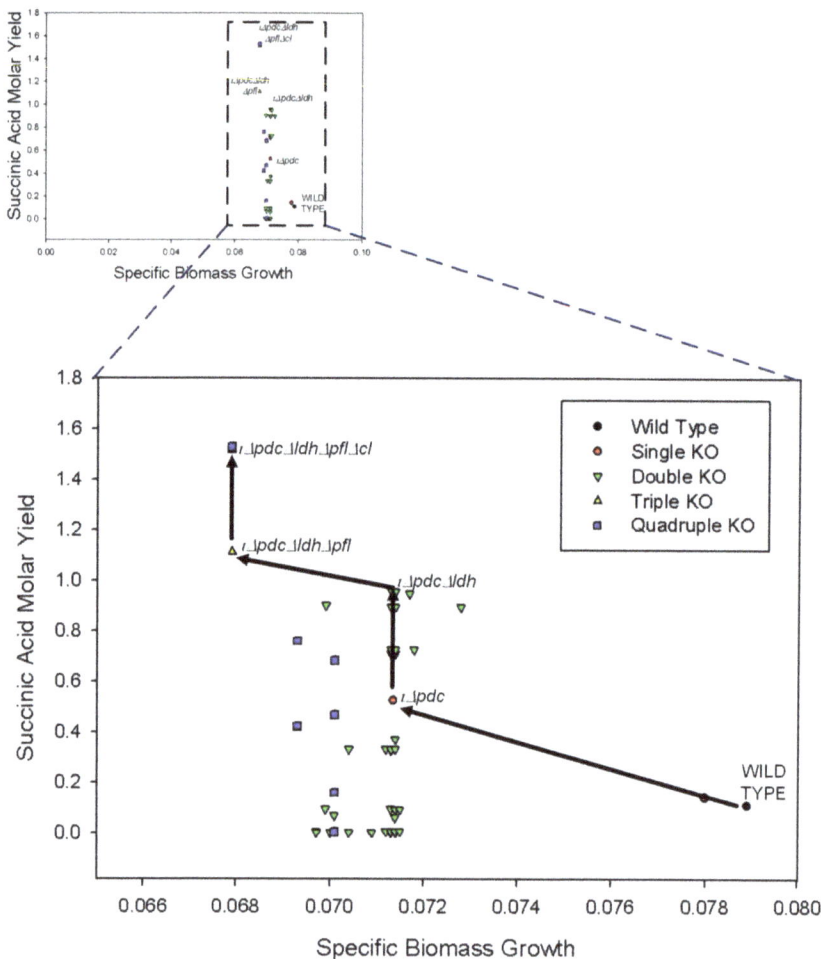

Figure 2. Simulation results for gene knockout analysis on central metabolic pathways. The shift in biomass growth and succinic acid molar yield from wild type to quadruple knockout (KO) is shown. The best strain for SA production is also shown for single (*i*Δ*pdc*), double (*i*Δ*pdc*Δ*ldh*), triple (*i*Δ*pdc*Δ*ldh*Δ*pfl*), and quadruple (*i*Δ*pdc*Δ*ldh*Δ*pfl*Δ*cl*) knockouts.

Figure 3 illustrates the redistribution of central metabolism fluxes with sequential knockouts of *pdc*, *ldh*, *pfl*, and *cl*, explaining how the strain can be genetically engineered to increase SA production. As we can see, this 4-gene knockout diverts carbon flux entirely from ethanol to SA sequentially through lactic acid (*pdc*), formic acid (*ldh*), acetic acid (*pfl*), and finally SA (*cl*). SA production is more expensive energetically, as it requires one extra mol of NADH compared to ethanol. While ethanol production needs one mol of NADH for pyruvate conversion, SA needs one for malate formation, and the other for menaquinone/menaquinol cycle. Figure 3 also shows how the NADH pool quantified by the flux-sum or turnover rate (see Materials and Methods) increases gradually, as ethanol is diverted to SA via lactic, formic, and acetic acids. Hong and Lee [56] have also observed in *E. coli* that higher SA production required more NADH. In contrast, ADP/ATP pool remains constant. It is also interesting to see that, while pyruvate is common to *pdc*, *ldh*, and *pfl*, only the knockout of *pdc* reduces its pool. This has a direct effect on growth, thus, the growth of *Z. mobilis* depends on the size of pyruvate pool. However, the additional knockouts of *ldh*, *pfl*, and *cl* do not affect the pyruvate pool, and hence have no effect on growth.

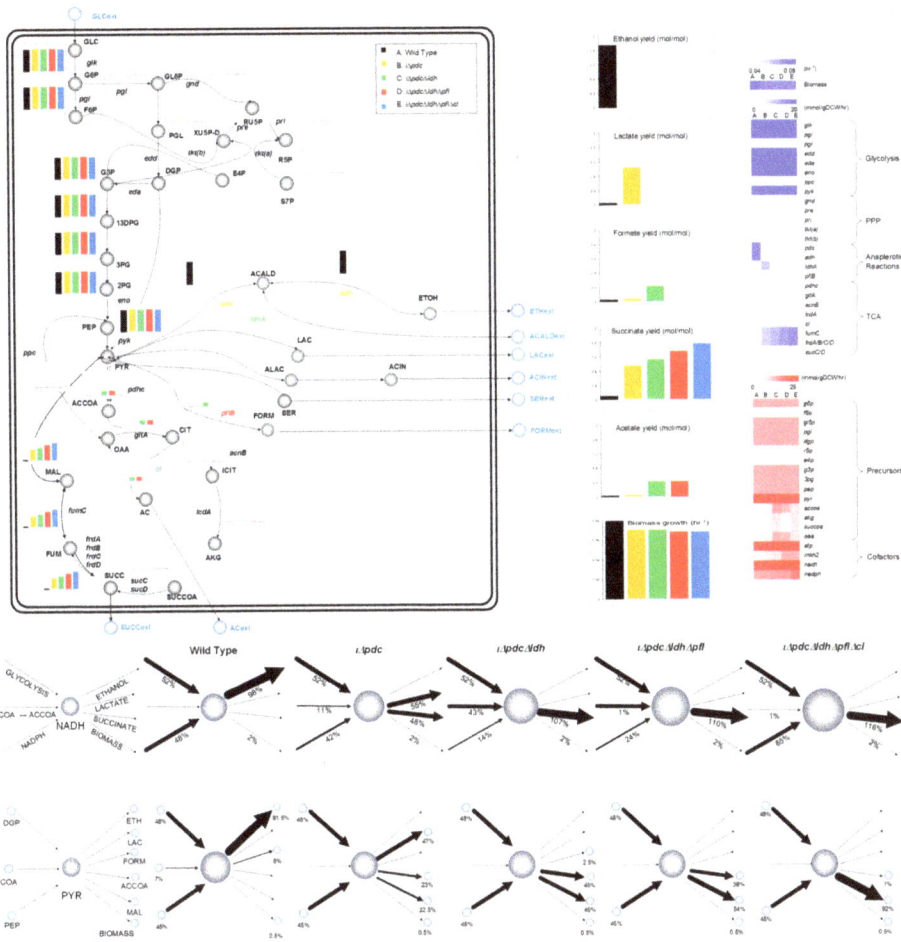

Figure 3. Metabolic flux distribution across the central metabolic pathways during the exponential growth phase of the microbial culture and flux-sum across the metabolite pyruvate and cofactor NADH in *Z. mobilis*. Consumption and production of the cofactor NADH and the metabolite pyruvate (PYR) is shown using the flux-sum values across each of the strains (wild type, *iΔpdc*, *iΔpdcΔldh*, *iΔpdcΔldhΔpfl*, *iΔpdcΔldhΔpflΔcl*). Percentage contribution is also shown. NADH: reduced nicotinamide adenine dinucleotide; NADPH: reduced nicotinamide adenine dinucleotide phosphate; COA: coenzyme A; ACCOA: acetyl coenzyme A; PYR: pyruvate; PEP: phosphoenolpyruvate; DGP: 2-dehydro-3-deoxy-D-gluconate 6-phosphate; MAL: malate.

3.4. Model-Driven Systematic Framework for Strain Optimization

Based on our analyses in the previous sections, we now propose a systematic model-driven framework for strain engineering to produce a desired target metabolite using a microbial host (Figure 4). Given a target metabolite, the first step of this framework is to select an expression host from the microbes that could potentially produce it. This expression host should have some special characteristics that would address some deficiencies such as low productivity in other expression hosts. In case of *Z. mobilis*, the high catabolic rate in anaerobic fermentation was the key feature that may offer a faster metabolite production rate. The second step is to develop and use a genome-scale model for the expression host to predict its maximum theoretical yield for the target metabolite. If the theoretical yield is comparable to those for other known producers, then the selected microbe has the potential to become a host for strain improvement.

The third step is to explore various combinations of gene modifications (deletions and/or additions) in the expression host to overproduce the target metabolite. This can be done in several ways. One way is to compare the metabolic network of the selected expression host with that of a known reference strain to identify the key differences in genes and pathways. Such a rational approach may give unique gene candidates in the reference strain and expression host for additions and deletions, respectively. The hypothesis is that the reference has optimally evolved and developed relevant synthetic pathways towards the desired metabolite [9]. In cases where only a few genes can be manipulated, as was the case in this work, this approach can be used to narrow down the pool of possible gene targets. For a small pool, it may be feasible to exhaustively consider all possible knockout combinations as done in the current study. These targets can then be evaluated using the genome-scale metabolic model to predict yields. However, for a large pool, such an exhaustive search may not be effective due to the combinatorial explosion. Therefore, another approach would be to exploit optimization techniques for strain improvement such as OptKnock [40], OptStrain [41], OptGene [42], OptForce [43], etc. While these methods may require some computational effort, some of them may guarantee the best solutions, if they converge. Both OptKnock and OptGene suggest the same strain modification as the one we have obtained in this work (Table 1). The multiple solutions of metabolites production yield observed are the results of different tools used. Note that, the SA yields obtained from different tools are almost consistent with the yields obtained from combinatorial knockout.

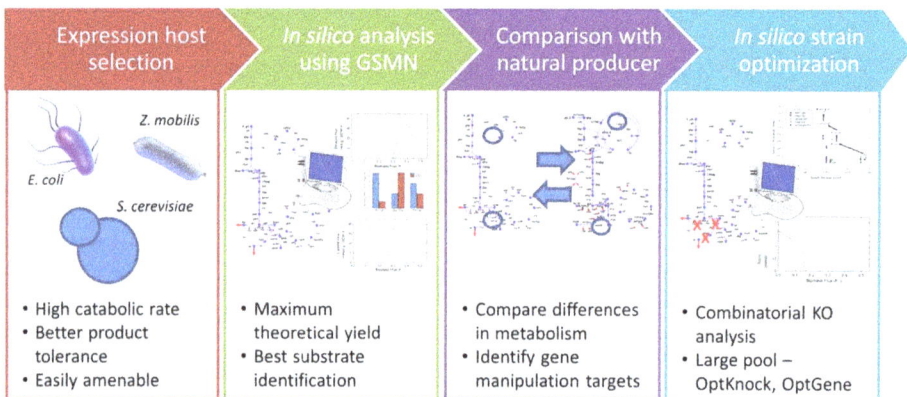

Figure 4. Proposed framework for strain improvement to overproduce a desirable target metabolite in a potential expression host (succinic acid in *Z. mobilis* in the current study).

4. Conclusions

In this study, we presented a systematic approach for overproducing SA in *Z. mobilis* by using an *in silico* constraints-based flux analysis. Our genome-scale model shows that *Z. mobilis* has the capability to produce SA. Our comparison of the central metabolisms of *Z. mobilis* and the SA-producer, *M. succiniciproducens*, suggested the removal of three genes (*pdc*, *als*, and *cl*) to mimic the metabolism of the latter. The inactivation of *pdc* and *cl* increased the SA yield, but the inactivation of *als* led to zero biomass growth. However, our combinatorial gene knockout analyses pointed to the inactivation of four genes *pdc*, *ldh*, *pfl*, and *cl*, which increased yield up to 1.5 mol SA per mol glucose. This is close to the maximum possible theoretical yield of 1.7 mol SA from glucose in *Z. mobilis*. Interestingly, *pdc*, *ldh*, and *pfl* are pyruvate-consuming reactions in *Z. mobilis*, whose sequential disruption seems to redistribute pyruvate flux from ethanol to lactic acid to formic acid to acetic acid, and then finally to succinic acid homofermentatively. Based on our work, we proposed a systematic framework to overproduce a desired metabolite from an organism by using the genome-scale metabolic model. The strain engineering strategies proposed in this work can be verified in the future through wet experiments.

Supplementary Materials: The supplements are available online at http://www.mdpi.com/2227-9717/6/4/30/s2, The supplements 1 and 2 contain complete updated model and biomass equations, respectively.

Acknowledgments: This work was supported by by the Academic Research Fund (R-279-000-476-112) of the National University of Singapore, Biomedical Research Council of A*STAR (Agency for Science, Technology and Research), Singapore and a grant from the Next-Generation BioGreen 21 Program (SSAC, No. PJ01334605), Rural Development Administration, Korea.

Author Contributions: H.W., I.A.K., and D.-Y.L. conceived and designed the *in silico* study; H.W. performed the computational simulations and drafted the manuscript; N.-R.L., and D.-Y.L. revised the paper.

Conflicts of Interest: The authors declare no conflict of interest.

References

1. Cukalovic, A.; Stevens, C.V. Feasibility of production methods for succinic acid derivatives: A marriage of renewable resources and chemical technology. *Biofuels Bioprod. Biorefining* **2008**, *2*, 505–529. [CrossRef]
2. Beauprez, J.J.; De Mey, M.; Soetaert, W.K. Microbial succinic acid production: Natural versus metabolic engineered producers. *Process Biochem.* **2010**, *45*, 1103–1114. [CrossRef]
3. Tan, T.; Liu, C.; Liu, L.; Zhang, K.; Zou, S.; Hong, J.; Zhang, M. Hydrogen sulfide formation as well as ethanol production in different media by cysND- and/or cysIJ-Inactivated mutant strains of Zymomonas mobilis ZM4. *Bioprocess Biosyst. Eng.* **2013**, *36*, 1363–1373. [CrossRef] [PubMed]
4. Jantama, K.; Haupt, M.J.; Svoronos, S.A.; Zhang, X.; Moore, J.C.; Shanmugam, K.T.; Ingram, L.O. Combining metabolic engineering and metabolic evolution to develop nonrecombinant strains of Escherichia coli C that produce succinate and malate. *Biotechnol. Bioeng.* **2008**, *99*, 1140–1153. [CrossRef] [PubMed]
5. Thakker, C.; Martínez, I.; San, K.-Y.; Bennett, G.N. Succinate production in Escherichia coli. *Biotechnol. J.* **2012**, *7*, 213–224. [CrossRef] [PubMed]
6. Zhu, L.W.; Xia, S.T.; Wei, L.N.; Li, H.M.; Yuan, Z.P.; Tang, Y.J. Enhancing succinic acid biosynthesis in Escherichia coli by engineering its global transcription factor, catabolite repressor/activator (Cra). *Sci. Rep.* **2016**, *6*. [CrossRef] [PubMed]
7. Wu, M.; Guan, Z.; Wang, Y.; Ma, J.; Wu, H.; Jiang, M. Efficient succinic acid production by engineered Escherichia coli using ammonia as neutralizer. *J. Chem. Technol. Biotechnol.* **2016**, *91*, 2412–2418. [CrossRef]
8. Mienda Faezah Mohd, B.S. Bio-succinic acid production: *Escherichia coli* strains design from genome-scale perspectives. *AIMS Bioeng.* **2017**, *4*, 418–430. [CrossRef]
9. Lee, S.; Lee, D.; Kim, T.; Kim, B. Metabolic engineering of Escherichia coli for enhanced production of succinic acid, based on genome comparison and in silico gene knockout simulation. *Appl. Environ. Microbiol.* **2005**, *71*, 7880–7887. [CrossRef] [PubMed]
10. Millard, C.S.; Chao, Y.P.; Liao, J.C.; Donnelly, M.I. Enhanced production of succinic acid by overexpression of phosphoenolpyruvate carboxylase in Escherichia coli. *Appl. Environ. Microbiol.* **1996**, *62*, 1808–1810. [PubMed]

11. Wang, X.; Gong, C.S.; Tsao, G.T. Bioconversion of fumaric acid to succinic acid by recombinant *E. coli*. *Appl. Biochem. Biotechnol.* **1998**, *70–72*, 919–928. [CrossRef] [PubMed]

12. Vemuri, G.N.; Eiteman, M.A.; Altman, E. Effects of growth mode and pyruvate carboxylase on succinic acid production by metabolically engineered strains of *Escherichia coli*. *Appl. Environ. Microbiol.* **2002**, *68*, 1715–1727. [CrossRef] [PubMed]

13. Wang, D.; Li, Q.; Mao, Y.; Xing, J.; Su, Z. High-level succinic acid production and yield by lactose-induced expression of phosphoenolpyruvate carboxylase in ptsG mutant *Escherichia coli*. *Appl. Microbiol. Biotechnol.* **2010**, *87*, 2025–2035. [CrossRef] [PubMed]

14. Hong, S.H.; Kim, J.S.; Lee, S.Y.; In, Y.H.; Choi, S.S.; Rih, J.K.; Kim, C.H.; Jeong, H.; Hur, C.G.; Kim, J.J. The genome sequence of the capnophilic rumen bacterium *Mannheimia succiniciproducens*. *Nat. Biotechnol.* **2004**, *22*, 1275–1281. [CrossRef] [PubMed]

15. Stols, L.; Donnelly, M.I. Production of succinic acid through overexpression of NAD+-dependent malic enzyme in an *Escherichia coli* mutant. *Appl. Environ. Microbiol.* **1997**, *63*, 2695–2701. [PubMed]

16. Lin, H.; Bennett, G.N.; San, K.Y. Fed-batch culture of a metabolically engineered *Escherichia coli* strain designed for high-level succinate production and yield under aerobic conditions. *Biotechnol. Bioeng.* **2005**, *90*, 775–779. [CrossRef] [PubMed]

17. Lin, H.; Bennett, G.N.; San, K.Y. Genetic reconstruction of the aerobic central metabolism in *Escherichia coli* for the absolute aerobic production of succinate. *Biotechnol. Bioeng.* **2005**, *89*, 148–156. [CrossRef] [PubMed]

18. Lin, H.; Bennett, G.N.; San, K.Y. Metabolic engineering of aerobic succinate production systems in *Escherichia coli* to improve process productivity and achieve the maximum theoretical succinate yield. *Metab. Eng.* **2005**, *7*, 116–127. [CrossRef] [PubMed]

19. Chayabutra, C.; Wu, J.; Ju, L.K. Succinic acid production with reduced by-product formation in the fermentation of *Anaerobiospirillum succiniciproducens* using glycerol as a carbon source. *Biotechnol. Bioeng.* **2001**, *72*, 41–48. [CrossRef]

20. Lee, P.C.; Lee, W.G.; Kwon, S.; Lee, S.Y.; Chang, H.N. Succinic acid production by *Anaerobiospirillum succiniciproducens*: Effects of the H_2/CO_2 supply and glucose concentration. *Enzyme Microb. Technol.* **1999**, *24*, 549–554. [CrossRef]

21. McKinlay, J.B.; Laivenieks, M.; Schindler, B.D.; McKinlay, A.A.; Siddaramappa, S.; Challacombe, J.F.; Lowry, S.R.; Clum, A.; Lapidus, A.L.; Burkhart, K.B.; et al. A genomic perspective on the potential of *Actinobacillus succinogenes* for industrial succinate production. *BMC Genom.* **2010**, *11*. [CrossRef] [PubMed]

22. Park, D.H.; Zeikus, J.G. Utilization of electrically reduced neutral red by *Actinobacillus succinogenes*: Physiological function of neutral red in membrane-driven fumarate reduction and energy conservation. *J. Bacteriol.* **1999**, *181*, 2403–2410. [PubMed]

23. Bradfield, M.F.A.; Mohagheghi, A.; Salvachúa, D.; Smith, H.; Black, B.A.; Dowe, N.; Beckham, G.T.; Nicol, W. Continuous succinic acid production by *Actinobacillus succinogenes* on xylose-enriched hydrolysate. *Biotechnol. Biofuels* **2015**, *8*, 181. [CrossRef] [PubMed]

24. Lee, S.J.; Song, H.; Lee, S.Y. Genome-based metabolic engineering of *Mannheimia succiniciproducens* for succinic acid production. *Appl. Environ. Microbiol.* **2006**, *72*, 1939–1948. [CrossRef] [PubMed]

25. Lee, J.W.; Yi, J.; Kim, T.Y.; Choi, S.; Ahn, J.H.; Song, H.; Lee, M.-H.; Lee, S.Y. Homo-succinic acid production by metabolically engineered *Mannheimia succiniciproducens*. *Metab. Eng.* **2016**, *38*, 409–417. [CrossRef] [PubMed]

26. Johns, M.R.; Greenfield, P.F.; Doelle, H.W. Byproducts from *Zymomonas mobilis*. In *Bioreactor Systems and Effects*; Springer: Berlin/Heidelberg, Germany, 1991; pp. 97–121, ISBN 978-3-540-47400-5.

27. Swings, J.; De Ley, J. The biology of Zymomonas. *Bacteriol. Rev.* **1977**, *41*, 1–46. [PubMed]

28. Beer, L.L.; Boyd, E.S.; Peters, J.W.; Posewitz, M.C. Engineering algae for biohydrogen and biofuel production. *Curr. Opin. Biotechnol.* **2009**, *20*, 264–271. [CrossRef] [PubMed]

29. Martín, H.G.; Ivanova, N.; Kunin, V.; Warnecke, F.; Barry, K.W.; McHardy, A.C.; Yeates, C.; He, S.; Salamov, A.A.; Szeto, E.; et al. Metagenomic analysis of two enhanced biological phosphorus removal (EBPR) sludge communities. *Nat. Biotechnol.* **2006**, *24*, 1263–1269. [CrossRef] [PubMed]

30. Park, J.H.; Lee, S.Y. Towards systems metabolic engineering of microorganisms for amino acid production. *Curr. Opin. Biotechnol.* **2008**, *19*, 454–460. [CrossRef] [PubMed]

31. Rossouw, D.; Næs, T.; Bauer, F.F. Linking gene regulation and the exo-metabolome: A comparative transcriptomics approach to identify genes that impact on the production of volatile aroma compounds in yeast. *BMC Genom.* **2008**, *9*. [CrossRef] [PubMed]

32. Wang, Q.Z.; Wu, C.Y.; Chen, T.; Chen, X.; Zhao, X.M. Integrating metabolomics into systems biology framework to exploit metabolic complexity: Strategies and applications in microorganisms. *Appl. Microbiol. Biotechnol.* **2006**, *70*, 151–161. [CrossRef] [PubMed]

33. Mienda, B.S.; Shamsir, M.S.; Illias, M.R. Model-assisted formate dehydrogenase-O (fdoH) gene knockout for enhanced succinate production in Escherichia coli from glucose and glycerol carbon sources. *J. Biomol. Struct. Dyn.* **2016**, *34*, 2305–2316. [CrossRef] [PubMed]

34. Khodayari, A.; Chowdhury, A.; Maranas, C.D. Succinate Overproduction: A Case Study of Computational Strain Design Using a Comprehensive Escherichia coli Kinetic Model. *Front. Bioeng. Biotechnol.* **2015**, *2*. [CrossRef] [PubMed]

35. Ren, S.; Zeng, B.; Qian, X. Adaptive bi-level programming for optimal gene knockouts for targeted overproduction under phenotypic constraints. *BMC Bioinform.* **2013**, *14*. [CrossRef] [PubMed]

36. Zhuang, K.; Yang, L.; Cluett, W.R.; Mahadevan, R. Dynamic strain scanning optimization: An efficient strain design strategy for balanced yield, titer, and productivity. DySScO strategy for strain design. *BMC Biotechnol.* **2013**, *13*. [CrossRef] [PubMed]

37. Patil, K.R.; Åkesson, M.; Nielsen, J. Use of genome-scale microbial models for metabolic engineering. *Curr. Opin. Biotechnol.* **2004**, *15*, 64–69. [CrossRef] [PubMed]

38. Blazeck, J.; Alper, H. Systems metabolic engineering: Genome-scale models and beyond. *Biotechnol. J.* **2010**, *5*, 647–659. [CrossRef] [PubMed]

39. Teusink, B.; Van Enckevort, F.H.J.; Francke, C.; Wiersma, A.; Wegkamp, A.; Smid, E.J.; Siezen, R.J. In silico reconstruction of the metabolic pathways of Lactobacillus plantarum: Comparing predictions of nutrient requirements with those from growth experiments. *Appl. Environ. Microbiol.* **2005**, *71*, 7253–7262. [CrossRef] [PubMed]

40. Burgard, A.P.; Pharkya, P.; Maranas, C.D. OptKnock: A Bilevel Programming Framework for Identifying Gene Knockout Strategies for Microbial Strain Optimization. *Biotechnol. Bioeng.* **2003**, *84*, 647–657. [CrossRef] [PubMed]

41. Pharkya, P.; Burgard, A.P.; Maranas, C.D. OptStrain: A computational framework for redesign of microbial production systems. *Genome Res.* **2004**, *14*, 2367–2376. [CrossRef] [PubMed]

42. Patil, K.R.; Rocha, I.; Förster, J.; Nielsen, J. Evolutionary programming as a platform for in silico metabolic engineering. *BMC Bioinform.* **2005**, *6*. [CrossRef]

43. Ranganathan, S.; Suthers, P.F.; Maranas, C.D. OptForce: An optimization procedure for identifying all genetic manipulations leading to targeted overproductions. *PLoS Comput. Biol.* **2010**, *6*. [CrossRef] [PubMed]

44. Chung, B.K.S.; Lee, D.Y. Flux-sum analysis: A metabolite-centric approach for understanding the metabolic network. *BMC Syst. Biol.* **2009**, *3*, 117. [CrossRef] [PubMed]

45. King, Z.A.; Feist, A.M. Optimizing Cofactor Specificity of Oxidoreductase Enzymes for the Generation of Microbial Production Strains—OptSwap. *Ind. Biotechnol.* **2013**, *9*, 236–246. [CrossRef]

46. Lakshmanan, M.; Chung, B.K.-S.; Liu, C.; Kim, S.-W.; Lee, D.-Y. Cofactor modification analysis: A computational framework to identify cofactor specificity engineering targets for strain improvement. *J. Bioinform. Comput. Biol.* **2013**, *11*, 1343006. [CrossRef] [PubMed]

47. Lakshmanan, M.; Lee, N.-R.; Lee, D.-Y. Genome-Scale Metabolic Modeling and In silico Strain Design of *Escherichia coli*. In *Systems Biology*; Wiley-VCH Verlag GmbH & Co. KGaA: Weinheim, Germany, 2017; pp. 109–137, ISBN 9783527696130.

48. Widiastuti, H.; Kim, J.Y.; Selvarasu, S.; Karimi, I.A.; Kim, H.; Seo, J.S.; Lee, D.Y. Genome-scale modeling and in silico analysis of ethanologenic bacteria *Zymomonas mobilis*. *Biotechnol. Bioeng.* **2011**, *108*, 655–665. [CrossRef] [PubMed]

49. Lee, D.Y.; Yun, H.; Park, S.; Lee, S.Y. MetaFluxNet: The management of metabolic reaction information and quantitative metabolic flux analysis. *Bioinformatics* **2003**, *19*, 2144–2146. [CrossRef] [PubMed]

50. McCarl, B.A.; Meeraus, A.; van der Eijk, P.; Bussieck, M.; Dirkse, S.; Steacy, P. McCarl GAMS User Guide (2008). Available online: http//gams.com/docs/document.htm (accessed on 14 February 2004).

51. Sang, Y.L.; Lee, D.Y.; Tae, Y.K. Systems biotechnology for strain improvement. *Trends Biotechnol.* **2005**, *23*, 349–358.

52. Seo, J.S.; Chong, H.; Park, H.S.; Yoon, K.O.; Jung, C.; Kim, J.J.; Hong, J.H.; Kim, H.; Kim, J.H.; Kil, J.; et al. The genome sequence of the ethanologenic bacterium Zymomonas mobil is ZM4. *Nat. Biotechnol.* **2005**, *23*, 63–68. [CrossRef] [PubMed]

53. DiMarco, A.A.; Romano, A.H. D-Glucose transport system of Zymomonas mobilis. *Appl. Environ. Microbiol.* **1985**, *49*, 151–157. [PubMed]

54. Lee, K.Y.; Park, J.M.; Kim, T.Y.; Yun, H.; Lee, S.Y. The genome-scale metabolic network analysis of Zymomonas mobilis ZM4 explains physiological features and suggests ethanol and succinic acid production strategies. *Microb. Cell Fact.* **2010**, *9*. [CrossRef] [PubMed]

55. Seo, J.-S.; Chong, H.-Y.; Kim, J.H.; Kim, J.-Y. Method for Mass Production of Primary Metabolites, Strain for Mass Production of Primary Metabolites, and Method for Preparation Thereof. U.S. Patent Application 20090162910 A1, 25 June 2009.

56. Hong, S.H.; Lee, S.Y. Importance of redox balance on the production of succinic acid by metabolically engineered Escherichia coli. *Appl. Microbiol. Biotechnol.* **2002**, *58*, 286–290. [CrossRef] [PubMed]

57. Meijer, S.; Nielsen, M.L.; Olsson, L.; Nielsen, J. Gene deletion of cytosolic ATP: Citrate lyase leads to altered organic acid production in aspergillus niger. *J. Ind. Microbiol. Biotechnol.* **2009**, *36*, 1275–1280. [CrossRef] [PubMed]

Review

Improving Bioenergy Crops through Dynamic Metabolic Modeling

Mojdeh Faraji and Eberhard O. Voit *

The Wallace H. Coulter Department of Biomedical Engineering, Georgia Institute of Technology and Emory University, 950 Atlantic Drive, Atlanta, GA 30332-2000, USA; Mojdeh@gatech.edu
* Correspondence: Eberhard.Voit@bme.gatech.edu

Received: 4 September 2017; Accepted: 7 October 2017; Published: 18 October 2017

Abstract: Enormous advances in genetics and metabolic engineering have made it possible, in principle, to create new plants and crops with improved yield through targeted molecular alterations. However, while the potential is beyond doubt, the actual implementation of envisioned new strains is often difficult, due to the diverse and complex nature of plants. Indeed, the intrinsic complexity of plants makes intuitive predictions difficult and often unreliable. The hope for overcoming this challenge is that methods of data mining and computational systems biology may become powerful enough that they could serve as beneficial tools for guiding future experimentation. In the first part of this article, we review the complexities of plants, as well as some of the mathematical and computational methods that have been used in the recent past to deepen our understanding of crops and their potential yield improvements. In the second part, we present a specific case study that indicates how robust models may be employed for crop improvements. This case study focuses on the biosynthesis of lignin in switchgrass (*Panicum virgatum*). Switchgrass is considered one of the most promising candidates for the second generation of bioenergy production, which does not use edible plant parts. Lignin is important in this context, because it impedes the use of cellulose in such inedible plant materials. The dynamic model offers a platform for investigating the pathway behavior in transgenic lines. In particular, it allows predictions of lignin content and composition in numerous genetic perturbation scenarios.

Keywords: biochemical systems theory; biofuel; lignin biosynthesis; optimization; plant metabolism; recalcitrance

---

1. Introduction

Crops have been cultivated, bred, and improved for thousands of years, and some successes have been astounding: a modern corn cob weighs between 1 and 1 ½ pounds, whereas its early predecessor, the ancient Latin American grass teosinte, had an average fruit weighing about 35 grams [1]. Achieving this 20-fold increase took about 8000 years.

In contrast to food crops, bioenergy crops have not been investigated for very long, if one ignores the burning of wood and other organic materials. Due to its relative youth, research on bioenergy crops had the immediate advantage of a rich body of genetic and metabolic information. Furthermore, millions of dollars in tax incentives and subsidies reflect the determination of many countries around the world to advance the use of sustainable bioenergy products, wean the world off its dependence from fossil fuels, and reduce greenhouse emissions. As a consequence, genetic and metabolic engineering have made the production of ethanol, butanol, and fatty acids from corn or sugar cane, competitive. As an example, in September 2014, three plants in Iowa and Kansas started commercial production of cellulosic ethanol with an annual ramp-up capacity of 80 million gallons. While impressive, this amount constitutes only a fraction of the federal call for 1.75 billion gallons per year in the U.S.

The low-hanging fruit of using sources like corn, is in direct competition with the supply of food products, and it has become today's challenge to produce "second-generation" bioenergy from other plant materials that do not cause ethical concerns. This new type of biofuel research is sometimes called "advanced", because it relies on lignocellulosic materials, which by and large, correspond to inedible plant parts, such as corn stover, pine bark, grasses, and wood chips. In parallel, algae have been studied extensively, but so far, with little economic success.

The challenges associated with exploiting these plant sources are manifold, but often converge to two overarching issues. First, the energy is stored less in concentrated, easily accessible sugars and more in woody substances that are difficult to ferment; we will discuss this aspect later. Second, plants are enormously complex, which is in part due to large numbers of constituents, and their interactions. As an example, some species of Spruce (*Picea*) are predicted to possess between 50,000 and 60,000 genes [2], which is between two and three times the number of human genes [3]. This large number of genes presumably reflects considerable redundancies, metabolic plasticity, and numerous stress response mechanisms, which are needed to compensate for the plant's lack of motility, and result in distinct differences to animal physiology [4,5]. Whereas the human metabolome library (HML) lists slightly more than 1000 different metabolites in humans [6], the number of metabolites in the plant kingdom is estimated to lie somewhere between 200,000 and 1,000,000 [7–9]; indeed, the width of this range alone indicates how little of plant metabolism we truly understand. Of course, plants also contain uncounted proteins and structural elements, which all contribute to their survival, but also impede attempts of targeted alterations. Collectively, these features render genetic and metabolic engineering of plants very challenging.

Nonetheless, new gene editing techniques, for instance, based on clustered regularly interspaced short palindromic repeats (CRISPR) and CRISPR-associated protein 9 (Cas9), have found their way into plant breeding [10]. The first applications targeted *Arabidopsis*, tobacco, rice, and wheat [11–13], but numerous other species have followed (e.g., [14–16]). These experimental advances are directly pertinent for future crop modeling, as they permit modifications that were considered impossible just a few years ago.

A particular, additional challenge associated with plants is that primary and secondary metabolism are tightly linked and regulated by "super-coordinated" gene expression networks [17]. This tight coordination may explain why it is not straightforward to identify and tweak only certain processes of interest, because many processes are possibly affected and robustly compensated. Adding to these complications is the fact that plant cells are highly compartmentalized [18,19], and that metabolite turnover occurs over a wide range of rates [20].

Another very challenging feature of plants is their polyploidy, that is, the existence of several copies of their chromosomes, which obviously makes targeted alterations cumbersome, and complicates essentially all gene manipulations, even if the methods and techniques are routine in microbes. Polyploidy is particularly important for crops, because it offers the opportunity of modifying traits and lineages [21]. In fact, polyploidization is found in many modern crops and wild species, including cotton (*Gossypium hirsutum*), tobacco (*Nicotiana tabacum*), wheat (*Triticum aestivum*), canola (*Brassica napus*), soybean (*Glycine max*), potato (*Solanum tuberosum*), and sugarcane (*Saccharum officinarum*). For instance, bananas are triploid and potatoes tetraploid, while wheat is hexaploid and sugarcane octoploid [22]. Polyploidy can be extreme: members of the genus *Ophioglossum* of adder's-tongue ferns can have very high chromosome counts, with possibly up to 720 chromosomes, due to polyploidy [23]. Further complicating the large numbers of chromosomes is the fact that some plants are alloploid, as they evolved or were bred through the hybridization of different species. An important example is the genus *Brassica*, which contains cabbages, as well as cauliflower, broccoli, turnip and seeds for the production of mustard and for canola oil (cf. [24]). Polyploidy can be traced back far within the phylogeny of a species. As just one example, there is strong evidence of polyploidy through breeding that can be seen in the long history of rice cultivation, where massive gene duplications have occurred since ancient times. The result is an estimated count of at least 38,000–40,000 genes in rice, of which only 2–3% are unique to any two rice subspecies, such as *indica* and *japonica* [25]. It is, at this point, unclear what the ramifications of polyploidy for modeling might be, but it is clear that both experimental and modeling approaches have to grapple with this key issue of crop manipulation.

Faced with the challenges and the enormous diversity of plants, the plant and crop communities have been focusing primarily on a number of model plants. Some of these, notably rice (*Oryza sativa*), maize (*Zea mays*), soybean (*Glycine max*), tobacco (*Nicotiana tabacum*), alfalfa (*Medicago truncatula*), and black cottonwood (*Populus trichocarpa*), are food, feed, or potential energy sources, while others, like *Arabidopsis thaliana* and *Brachypodium distachyon*, have features that greatly facilitate their investigation, such as relatively small genomes, fast growth, and diploidy instead of polyploidy.

Within this context, most improvements in crop production have come from experimental metabolic engineering research, which has been model free in the sense that new plant alterations were guided by biological intuition. This approach has been very successful in many instances, but one should expect it to run into problems as soon as large omics datasets become a standard in crop science. These datasets are very valuable, but they are also so immense in size that they cannot be comprehended by the unaided human mind, and require sophisticated computer algorithms for analysis and interpretation; a review by Yuan et al. [5] discusses this need for integrating "big data" with traditional plant systems biology. Yet, even modern machine learning methods of analysis are not sufficient by themselves. These methods are designed to filter information from noise and mine patterns from data that the unaided human mind cannot comprehend, but they rarely suggest mechanisms or provide explanations, especially with respect to the dynamics of a system under study, and if this system is nonlinear, due to regulation, synergisms, and threshold effects. Thus, if the goal of an investigation is an explanation of why a system behaves the way it does, or a prediction of system responses under untested conditions, intuition and statistical data analysis alone are susceptible to failure in the complex world of plant physiology. They need to be complemented with dynamic systems modeling, which is a rather new subject in bioenergy research.

Among the relatively few recent mathematical modeling efforts in the field of crop research, many studies have focused on photosynthesis (e.g., [26–33]), on pathways of general importance, such as the TCA cycle (e.g., [4,31,34–36]), or on specific pathways that are of another industrial interest, such as flavonoid and isoprenoid metabolism (see reviews [5,19,37,38]). By contrast, metabolic modeling for improved plant biofuels is still relatively scarce. Nonetheless, considering the complexity and variability of plants, the utilization of methods of computational systems biology appears to become an increasingly rational strategy toward realizing economically feasible bioenergy production, and one might expect that computational modeling will become a standard tool of guiding experimentation in the future.

Returning to the specific challenge of difficult access to sugars in inedible plant materials, the focus must shift to lignin, which severely impedes bioenergy extraction from woody substrates. Except for cellulose, lignin is the most abundant terrestrial biopolymer and accounts for roughly 30% of all organic carbon in the biosphere [39]. It is the main constituent of wood, and plays a vital role in terrestrial plant life, as it is the key component of the water transport system in the plant xylem, and gives the plant structure and strength, to overcome gravity. Chemically, lignin is an irregular phenolic polymer, whose hydrophobic nature not only facilitates water transfer from the roots, but also blocks surface evaporation from stems and leaves.

Within the cell wall, lignin is physically entangled with cellulose and hemicellulose molecules, and thereby, severely limits the production of ethanol and other bioenergy compounds by hindering the access of enzymes to these desirable polysaccharides. It is also resistant to enzymatic digestion, and therefore, difficult to remove from plant materials. The ultimate consequence of lignin in plant walls is *recalcitrance*, which summarily describes the resistance of plant materials to fermentation. Recalcitrance has emerged as a major obstacle in the commercial production of cellulosic ethanol and other bioenergy compounds, and has therefore become a key target for bioenergy research. The complete elimination of lignin is, of course, not desirable, but even a reduction in lignin content and/or certain changes in lignin composition have been shown to improve ethanol yield [40–43]. As a consequence, the second-generation bioenergy industry has put lignin biosynthesis and degradation into the spotlight. Specifically, one focus area has become the alteration of the quantities and proportions of the three or more types of monolignols, which are the building blocks of the lignin heteropolymer.

As an interesting side note, lignin is not always a problem. In fact, it is a true yin and yang: on the one hand, it is an impediment to bioenergy production, but on the other hand, is a very intriguing organic compound, and some recent industry efforts actually target the harvesting of lignin as a valuable resource for a variety of chemical syntheses.

Most efforts of altering lignin have been directed toward biotechnological experimentation, and computational modeling efforts are still the exception, although they have emerged with increasing frequency. Examples include computational models by Lee et al., who analyzed the pathways of lignin biosynthesis in poplar [44] and alfalfa [45], based on gene knockdown experiments. Wang et al. constructed a kinetic model of the lignin pathway from a large set of in vivo and in vitro measurements [46]. Faraji et al. investigated the lignin biosynthesis pathway in switchgrass (*Panicum virgatum*) [47], which was identified by the U.S. Department of Energy as the most promising monocot plant for biofuel ethanol by DOE. Other works in this field include [48,49]. A summary of highlights from these studies is provided in [50].

In the following, we first describe representative mathematical approaches that are currently used for modeling crop metabolism, and include a wide variety of techniques. One should note that the physiological attributes of plants often translate into unique mathematical constraints that require reevaluation of the details and underlying principles of popular modeling formalisms. While discussing different approaches, we highlight specifically the metabolic modeling of bioenergy crops. Subsequently, we finish with a modeling case study that explores the pathway of lignin biosynthesis in switchgrass. In terms of references, we give preference to articles addressing plant and crop systems, while keeping general references to a minimum.

2. Mathematical Modeling Approaches for Metabolic Engineering in Crops

Significant improvements in food and bioenergy crops are very challenging, but the enormous global scale of mobile energy use, and the corresponding potential economic benefits of even minor percent improvements in biofuel yield, are very attractive. As a consequence, many attempts have been made to alter crops with traditional methods of metabolic engineering, where the overriding goal is the targeted alteration of metabolic pathways toward better yields in compounds, like ethanol and butanol.

It is only recent that computational biology has begun to partner with experimental biology in advancing and pushing the boundaries of rational crop science [51]. Much of this work has focused on the model plant species discussed earlier, and indeed, some comprehensive, multi-scale models are available that address these model species [52–55]. Also, most of these models have an exclusive focus on steady-state operation, whereas dynamic modeling of bioenergy pathways is still in its infancy. Baghalian et al. [37], and Morgan and Rhodes [19], provide excellent reviews on modeling plant metabolism that describe prominent mathematical approaches in the field.

Mathematical models for metabolic pathway analysis are manifold, and driven by the availability of data types [56]. They may be classified in two coarse categories. The first uses steady-state approaches, which have the two advantages that they are algebraic, which renders large model sizes possible, and that they are relevant, as many systems operate close to a steady state. The second category contains dynamic models, which are more realistic, and cover transients as well as the steady state, but are mathematically more complicated. Outside these categories, the literature contains a few models that are stochastic, permit spatial considerations, and span multiple scales.

2.1. Steady-State Modeling

Any modeling strategy is ruled by the availability of data, and plant metabolic modeling is no exception. For systems operating close to a steady state, ideal data would consist of metabolite concentrations, and of the distribution of fluxes throughout a metabolic system. Unfortunately, such data are seldom available, and the computational estimation of fluxes has become one of the crucial steps in metabolic modeling. One basis for estimation is the technique of ^{13}C labeling, which has become popular for metabolic flux characterizations, and entails computational modeling for metabolic network reconstruction [57,58]. In particular, stoichiometric modeling [59] has been widely applied to isotopic labeling data [36,60–64]. In some cases, these methods have allowed the estimation of entire flux maps [18], but most metabolic flux models presently lack sufficient data and are underdetermined. As a consequence, much effort in the field has been dedicated to algorithm development and experimental techniques that try to infer flux values from other data.

The estimation of fluxes falls into two steps. First, one needs to determine which fluxes are likely to exist within a particular metabolic system. This determination is usually performed indirectly, namely through genome sequencing, which permits connecting a genotype to an observable phenotype by means of genome-wide metabolic reconstructions, based on sequence comparisons with better-identified organisms [65,66]. Once the candidate fluxes and their associations with metabolites are established, the magnitudes of all fluxes are to be determined. The guiding principle is that, at any steady state, the fluxes entering a metabolite pool must collectively be equal, in total magnitude, to the collection of fluxes exiting this pool [67,68]. The most prominent implementation of this concept is flux balance analysis (FBA; see below) [65,69], which computes the flux distribution within a metabolic pathway at a steady state, based on an assumed objective of the system, such as maximum growth or some maximum flux.

Other steady-state approaches at the level of flux distributions are flux variability analysis (FVA) [70], elementary mode analysis (EMA) [71,72], extreme pathway analysis (EPA) [71,72], and metabolic flux analysis (MFA) [73]. One might also mention metabolic control analysis (MCA) [74–77] in this category, as it was designed specifically for assessing the control of flux through a pathway at a steady state. Pertinent details of these approaches are presented below.

2.1.1. Flux Balance Analysis (FBA)

In typical pathway systems, the number of fluxes is greater than the number of metabolites, because the same metabolite is usually involved in more than one reaction. As a consequence, the stoichiometric matrix of a typical metabolic system is underdetermined and infinitely many solutions are possible. To address this situation, FBA formulates the system as a linear programing problem, where the solution of the underdetermined system is a member of the solution space, and optimizes an objective function of choice, such as maximal growth. The solution space itself is determined by linear constraints of the problem, such as non-negativity and maximal magnitudes of fluxes [64,78]. FBA is a simple, yet powerful tool that has been widely used to determine steady-state flux distributions. One caveat of this method is the choice of a suitable objective function. The choice of maximal growth is often suited for microbial populations, but in mammalian systems and in plants, where several pathways simultaneously share metabolites and enzymes, selecting the right objective function is not a straightforward task.

FBA has been used successfully in plant and bioenergy research. For instance, Paez et al. analyzed biomass synthesis in *Chlamydomonas reinhardtii* under different CO_2 levels [79]. Chang et al. presented a genome-scale metabolic network model of the same organism [80] using FBA and FVA. Employing a variant of flux balance analysis that accounts for dynamics (DFBA), Flassig et al. [81] modeled the β-carotene accumulation in *Dunaliella salina* under various light and nutrient conditions. Because it is to be expected that plants must satisfy several objectives, methods of multi-objective optimization have been applied to metabolic plant modeling as an alternative to FBA [82,83].

A somewhat problematic aspect of FBA is the omission of nonlinearities, such as regulatory signals, which clearly operate in actual cells. While the FBA solution itself is unaffected by regulation, any extrapolations to new situations, such as gene knockouts, can be significantly influenced by regulatory signals, thereby rendering FBA predictions questionable. A second issue is the fact that plant cells are highly compartmentalized, which complicates any type of modeling. In particular, one must question whether it is admissible to merge "parallel" fluxes, using the same substrates, which are, however, proceeding in different compartments. Experimental studies have shown that even within the cytosol, spatial channeling of multiple enzymes can mimic pseudo-compartmental behavior, without which, some aspects of the dynamics of a plant cell cannot be explained [84]. Finally, it is unclear to what degree plant cells truly operate under (quasi-) steady-state conditions.

An interesting variation of FBA is the *method of minimization of metabolic adjustment* (MOMA) [85], which characterizes a flux distribution that is altered due to a mutation or intervention, in relation to the corresponding FBA solution for the same wild type internode. Expressed differently, MOMA focuses on the admissible solution within the solution simplex that most closely mimics the wild type. Lee and others used MOMA to analyze data from knockdown experiments with genes associated with lignin biosynthesis in alfalfa [45].

2.1.2. Flux Variability Analysis (FVA)

FVA is a constraint-based modeling variant of FBA. It addresses the well-known situation in linear programming (LP), that a problem has infinitely many solutions, because the optimal solution is not one of the vertices of the solution simplex. This situation can arise, for instance, when the objective function is parallel to one of the LP constraints. For such a case, FVA determines the variability in each of the fluxes in the proximity of equivalent, admissible solutions [70]. An unexpected merit of the method is that biological systems do not necessarily operate truly optimally, and that it is hence important to explore flux distributions in slightly suboptimal solutions as well. As a more conservative method, FVA may appear to be a better fit for the complex biology of plants than FBA.

Hay and Schwender employed flux variability analysis (FVA) to reconstruct the seed storage metabolism pathway in oilseed rape (canola; *Brassica napus*), and to characterize the changes in this pathway during seed development [86,87]. As one of the largest sources of edible vegetable oil in the world, oilseed rape is also a favored biofuel crop. The authors were able to identify the differential roles of fluxes and their variability under different nutritional conditions. Their results provide an interesting computational validation of how metabolic redundancies can play a crucial role during the important phase of seed development with the rapeseed life cycle.

In a different application of FVA, combined with FBA, Chang et al. developed a genome-scale metabolic network model for the microalga *Chlamydomonas reinhardtii* [80], to investigate the effect of light on metabolism. Their specific goal was to create a predictive tool for an optimal light source design.

2.1.3. Extreme Pathway Analysis (EPA) and Elementary Mode Analysis (EMA)

Extreme pathways represent the structure of a pathway network as a linear combination of flux pathways that act as the vector basis, in the sense of linear algebra [71,72]. With this set-up, any steady-state vector of the system can be written as a linear combination of this basis. In a geometric interpretation, the extreme pathways are the lateral edges of the admissible cone of solutions that is anchored at the origin. The extreme pathways are a subset of the so-called elementary modes of the pathway system. In EMA, non-decomposability constraints ensure these elementary modes are genetically independent. As a consequence, they can explain the links between the genotypes and the corresponding phenotypes. Steuer et al. applied elementary modes in their analysis of the mitochondrial TCA cycle in plants [88].

Extreme pathways are unique and irreducible sets of elementary modes. As such, an important drawback arises when a pathway has many degrees of freedom, because the number of the elementary modes is equal to the degrees of freedom in the pathway. In such cases, analysis of the system through the assessment of elementary modes becomes cumbersome due to the combinatorial explosion of admissible routes in the system. [89]. A second limitation of the method is that extreme pathways cannot always convert an input into a desired product, although the elementary modes of the system allow such a conversion [72]. Also, typical EPAs assume a predominant reaction for every reaction, which is often, but not always a given. If the system contains reversible pathways, the extreme rays of the solution cone may lose this property, and it may be preferable to work with *extreme currents*, or to define different classes for reversible and irreversible fluxes [90].

The main advantage of EMA/EPA is the following: in a metabolic engineering problem, diagnostics of elementary modes and extreme pathways will assist in designing a scheme of multiple genetic alterations in a targeted manner. Specifically, an optimized solution derived from techniques such as linear programing, might provide a more desirable numerical value for the objective of the problem when suboptimal solutions derived from EMA provide more biologically meaningful solutions, due to the synergy in the regulation of the genes involved in the chain of reactions in each elementary mode.

2.1.4. Metabolic Flux Analysis (MFA)

MFA relies on labeling data, which are usually generated with an experiment where ^{13}C labeled substrate is given to the system. After some while, the label distributes among the metabolites according to the magnitudes of fluxes within the system. ^{13}C is a stable isotope of carbon which contains one extra neutron relative to ^{12}C, the most abundant isotope of carbon. Hence, methods such as mass spectrometry are able to detect the level of isotope abundance in different metabolites, which in turn, assists in the elucidation of the fluxes in a metabolic pathway. The idea behind MFA is that measuring sufficiently many fluxes leads to a substantial reduction of the degrees of freedom of a pathway, possibly to zero, in which case, a unique solution is achievable [72]. Although conceptually straightforward, MFA is technically quite difficult, and measuring internal fluxes is still a challenge. However, new experimental techniques are expected to provide us with the desired information in the foreseeable future [91]. Roscher et al. discussed applications of metabolic flux analysis in photosynthetic and non-photosynthetic plant tissues [92]. The comprehensive review by Dieuaide-Noubhani and Alonso [93] covers MFA in plants, and describes both experimental and mathematical modeling steps.

2.1.5. Metabolic Control Analysis (MCA)

MCA [74–77] was proposed specifically for assessing the control of flux through a pathway at a steady state. Before MCA was developed, it was assumed that every pathway has a rate-limiting step, which controls the flux through the pathway. The proponents of MCA showed convincingly that there is seldom a single rate-limiting step in a metabolic pathway. Instead, the control of the flux is distributed, with different degrees of importance, among many or all reaction steps. MCA addresses this issue by computing flux and metabolite control coefficients, and elasticity coefficients, which coarsely correspond to sensitivities, and may be derived from alleged functional forms of rate laws or direct experimental measurements [94]. A review by Rees and Hill discusses MCA specifically in the context of plant metabolism [95]. Giersch et al. [96] applied MCA to the system of photosynthetic carbon fixation.

2.1.6. Limitations of Steady-State Approaches

Steady-state modeling has the advantage of relative mathematical simplicity, and in particular, the fact that no differential equations are involved. However, the restriction to steady-state operation must be considered with some caution in plant and crop modeling, as plants seldom truly operate at the same steady state throughout the day [19,20,37,92]. In particular, the dynamics of the light–dark cycle is an important reminder of the non-steady-state operation of plants [36]. Parallel pathways, often occurring in several compartments, add to the complexity of plant metabolism [18]. Finally, the large range of turnover times of metabolite pools may affect the validity of pure steady-state models [27].

2.2. Dynamic Modeling

Dynamic modeling has the potential of capturing the complex physiology of plants more accurately. Specially, kinetic models are, at least in principle, capable of simulating time course data and permit a variety of dynamical analyses of metabolic pathways. However, in comparison to steady-state models, dynamic models are more difficult to analyze, and require much more data support.

2.2.1. Explicit Kinetic Models

The law of mass action is the basis for the earliest quantitative modeling of a chemical reaction rate law [97]. Kinetic mass action models are widely used in metabolic modeling. In plant metabolic modeling, an example is the work by Farre et al. [98], who developed a model of carotenoid biosynthesis in maize to identify effective genetic intervention points. Bai et al. [99] used a mass action kinetic model to investigate the carotenoid pathway in rice embryonic callus, and presented model-driven metabolic engineering strategies.

Rooted in the law of mass action, the first mechanistic kinetic models of metabolism were based on the concept of the Henri–Michaelis–Menten mechanism and its generalizations [100,101]. The mathematical representations of the reaction steps, according to these concepts, contain physical properties that are represented by measurable parameters, such as V_{max}, K_M, and K_i [100,102]. Although these mechanistic kinetic models are still predominant [46], their underlying assumptions are seldom justified in a living cell. For instance, these models implicitly rely on the homogeneity of the medium in which the reactions occur, which doesn't hold true in the in vivo environment of a cell. For larger systems, the parameterization of mechanistic kinetic models becomes laborious, expensive, and time consuming [37], and the resulting measurements are often obtained in vitro, and may not be representative of enzyme kinetics in vivo [27,103]. An additional problem with mechanistic models is that it may become impossible to infer their structure if multiple regulatory mechanisms are involved in a reaction [104].

The simplest mechanistic models of metabolism are over a century old, and describe the kinetics of individual enzyme catalyzed reactions [100,101,105]. A modern example of their use within the context of bioenergy crops is the work of Nag et al. [106], which elucidates the carbon flow in plant cells, using a mechanistic kinetic model of starch degradation. Starch is of great interest in the biofuel industry, as it is readily fermentable into alcohol or other energy products. Wang et al. [46] constructed a kinetic metabolic model of lignin synthesis in black cottonwood (*Populus trichocarpa*), based on a large array of in vivo and in vitro measurements.

Uncounted variations and alternatives of the original Michaelis–Menten concept were developed over the years, to represent more complicated enzymatic processes and their regulation in an appropriate manner. These variants have been reviewed numerous times [104,107,108], and are not described here much. Instead, we focus on more global approaches that permit streamlined representations of entire pathway systems. As alternatives to the original mechanistic models, several modeling frameworks have been proposed as semi-mechanistic strategies that represent which variables affect which fluxes, but do not dictate specific mechanisms. Examples include biochemical systems theory (BST) [44,45,47,109], structural kinetic modeling [110], dynamic flux estimation [111] and nonparametric dynamic modeling [112,113]. They are briefly described below.

2.2.2. Biochemical Systems Theory (BST)

BST is a kinetic modeling approach that uses power-law functions to model all fluxes [114–116]. The core idea behind BST is that, in logarithmic space and close to an operating point, a rate law is well represented by a linear function of the substrate(s) and regulator(s) of the reaction [117]. Therefore, a Taylor linearization in logarithmic space about the biological operating point approximates the often unknown kinetic process with reasonable accuracy. In Cartesian space, the result of this approximation is a term consisting of a rate constant, and a product of power-law functions of all contributing variables, each raised to an exponent, called the kinetic order. Each exponent can have any real value; it is positive for substrates and activators, negative for inhibitors, and zero for variables without direct effect on the flux. The power-law functions are easy to adjust for any number of substrates or regulators, and BST has been widely used in a variety of organisms. Lee et al. used BST in a steady-state and dynamic flux characterization of the lignin biosynthesis pathway in *Medicago* [109]. Other examples of BST framework in plant modeling are [44,45,47,118–125].

2.2.3. Other Dynamic Modeling Approaches in a Predefined Format

The *saturable and cooperative formalism* has its roots in BST, but instead of presenting the variables in each term with simple power-law functions, uses Hill-type functions [126]. Thus, every process is guaranteed to saturate, and the accuracy of models in this formalism is often higher than in BST. However, this improvement is paid for with a considerable increase in the number of parameters.

The *linear-logarithmic* (lin-log) representation of enzyme-catalyzed reactions is closely aligned with the concepts of MCA, and can be seen as the dynamic arm of this formalism [127,128]. It represents all variables, reaction rates, and fluxes in relation to their steady-state analogues. For very large substrate concentrations, the accuracy of these models is superior to those in BST [129], but for very small concentrations, the lin-log rates become negative and approach $-\infty$ when the substrate converges to zero [117,130,131].

Structural kinetic modeling recognizes the disadvantages of rate laws whose mathematical formats cannot be justified on biological grounds, and assigns a local linear representation at each point of a simulation. The system is first written in terms of the Jacobian of the system, and then the Jacobian is reconstructed such that its components are either directly measurable or estimated. The resulting model is free of explicit functional forms [110]. Steuer et al. presented a structural kinetic model of Calvin cycle in chloroplast stroma [110].

Dynamic flux estimation (DFE). Given the co-existence of very many, very different mathematical representations for metabolic processes, and the fact that none of these are mathematically guaranteed to be correct, with the exception of a small range of validity about an operating point, one might ask whether one can obtain a glimpse of true representations directly from data. A related question is whether it is absolutely necessary to specify functional formats before one starts modeling. DFE offers some answers to these questions.

DFE is a dynamic modeling approach that requires good time series of metabolite concentrations, and uses these to circumvent the initial need for selecting suitable functional forms and their parameterizations [111]. DFE does this by algebraically isolating each flux, and deriving graphical and numerical flux–substrate relationships, in the following manner. First, the time series data are smoothed, and the slopes of the time courses are numerically estimated at many time points. These slope values are substituted for the derivatives on the left-hand sides of the differential equations of the system. The result is a large algebraic system of equations that represent the pathway at many time points. This system is solved, and the result is a set of arrays that assign flux values to time points or to metabolites on which they depend. These results can be plotted, which reveals flux representations that are presumably very close to the truth. In an independent second step, one attempts to represent the numerical flux representations with parametric functions. In addition to being close to the truth, DFE minimizes compensation errors that commonly arise in a simultaneous parametrization of systems. A drawback of DFE is that its direct implementation would require a square stoichiometric matrix. However, several procedures have been proposed [113,132–135] to relax this assumption, which is seldom true.

Nonparametric dynamic modeling. When it comes to selecting the functional format of rate laws in DFE, there is no silver bullet. Whether mechanistic or non-mechanistic, any rate law needs to be mathematically specified, and then parameterized. A rare exception is the recently proposed method of nonparametric dynamic modeling, which circumvents the need to select functional forms by deriving and utilizing their shapes directly from time series data [112]. This method is a direct variant of DFE that uses the same initial steps, but then replaces the choice and fitting of functional forms with look-up tables that were derived from the data.

Specifically, the look-up tables are assembled from the flux–substrate relationships that were established in the first phase of DFE. The information in these look-up tables consists of discrete points on curves or surfaces representing flux values throughout the ranges of the experimental time series data. The numerical solver for the otherwise typical ODEs is discretized, and uses the look-up tables instead the closed-form rate laws to calculate flux values at each iteration. Because the method depends so strongly on available data, it is numerically valid only over the given experimental ranges and close-by. Although the nonparametric character of the method might appear to be a limiting factor, this type of dynamic modeling is surprisingly accurate and powerful. For instance, it is possible to perform stability and sensitivity analysis, and to compute steady states from non-steady-state data. By its definition, nonparametric dynamic modeling is an essentially unbiased approach that is almost free of assumptions.

2.3. Other Approaches: Stochastic, Spatial, and Multi-Scale Models

Stochastic models. Models containing randomness have been studied for a long time within the realm of statistical analyses of stochastic processes. In the context of plants and crops, Hartmann and Schreiber analyzed sucrose degradation using various formalisms, including stochastic Petri net (SPN) simulations in potato (*Solanum tuberosum*) [136]. Wu and Tian developed a stochastic multistep modeling framework to improve the accuracy of delayed reactions [137], and applied their method to the aliphatic glucosinolate biosynthesis pathway. One notable phenomenon in plants is the circadian clock, which has been studied in detail in the bread mold *Neurospora crassa* [138,139]. The review by Guerriero et al. [140] presents examples of stochastic models that investigate the effects of intrinsic noise in these circadian rhythms (see also [141–143]).

Spatial models. The importance of spatial assumptions in a plant metabolic model was earlier discussed. To capture the highly compartmentalized environment of plant cells demands a more detailed approach than is possible with the models discussed so far. A good example in this category is the work by Bogart and Myers [53], who constructed a spatial model of a maize leave to explain its metabolic state in response to a developmental gradient observed between base and tip tissue. The review by Sweetlove and Fernie [144] gives an overview of spatial modeling in plants, and identifies experimental and computational challenges that must be overcome before a realistic spatial or compartmental model for plants can be achieved.

Multi-scale models. Multi-scale modeling is challenging because different aspects of a system in space, time and biological organization require different degrees of granularity. For instance, a model that includes a wide range of time scales suffers from necessary compromises between high temporal resolution in detailed modules, and coarseness at a higher level; it may also have problems with stiffness. By contrast, a model focusing on a very narrow time scale might not capture the essence of a system's dynamics. Therefore, the ideal may be a hierarchical hybrid modeling scheme, with an ensemble of modules where each module covers a certain time scale, which often aligns with corresponding spatial and organizational scales.

In spite of these challenges, prominent examples of multi-scale modeling have been elaborated in the form of multi-organ, and even whole-plant models. A multi-organ FBA model by Grafahrend-Belau et al. [145], was developed to investigate the metabolic behavior of source and sink organs during the generative phase in barley (*Hordeum vulgare*). The SOYSIM project is a whole-plant model developed by University of Nebraska at Lincoln, that simulates the soybean growth from emergence to maturity [55]. WIMOVAC is a simulation model of vegetation responses to environmental changes; it focuses specifically on the carbon balance in plants [54].

2.4. Case Study: Lignin Biosynthesis in Switchgrass

As discussed earlier, an intermediate goal of improving bioenergy yields is the alteration of lignin toward reduced recalcitrance, which necessitates a solid understanding of how monolignols are synthesized before they are sent from the cytosol into the cell wall. Achieving this goal requires good experimental data regarding monolignol biosynthesis, and a computational framework for the effective analysis of these data. The latter is the subject of this case study describing analyses of lignin biosynthesis in switchgrass (*Panicum virgatum*).

Four different monolignols are the major precursors of the lignin polymer in *P. virgatum*. They are called *p*-coumaryl alcohol, coniferyl alcohol, 5-OH-coniferyl alcohol, and sinapyl alcohol, or, more casually, H-lignin, G-lignin, 5H-lignin, and S-lignin, respectively. The lignin pathway in switchgrass is a subset of the phenylpropanoid pathway, and has a topology that is distinct from other plants, and uses a particular set of enzymes. Predictions regarding the dynamics of the pathway are difficult, because several of these enzymes (4CL, CAD, CCR, COMT, F5H) are shared among multiple reactions with different substrates (Figure 1). Specifically, the part of the pathway leading to S- and G-lignin has a grid-like structure. The shared enzymes, the grid structure, and the flux representations are the foundation of the nonlinear characteristics of the pathway, and make intuitive predictions regarding perturbations or interventions unreliable. In particular, it is difficult to foresee how up- or downregulation of genes would affect the composition of monolignols under non-wild type conditions.

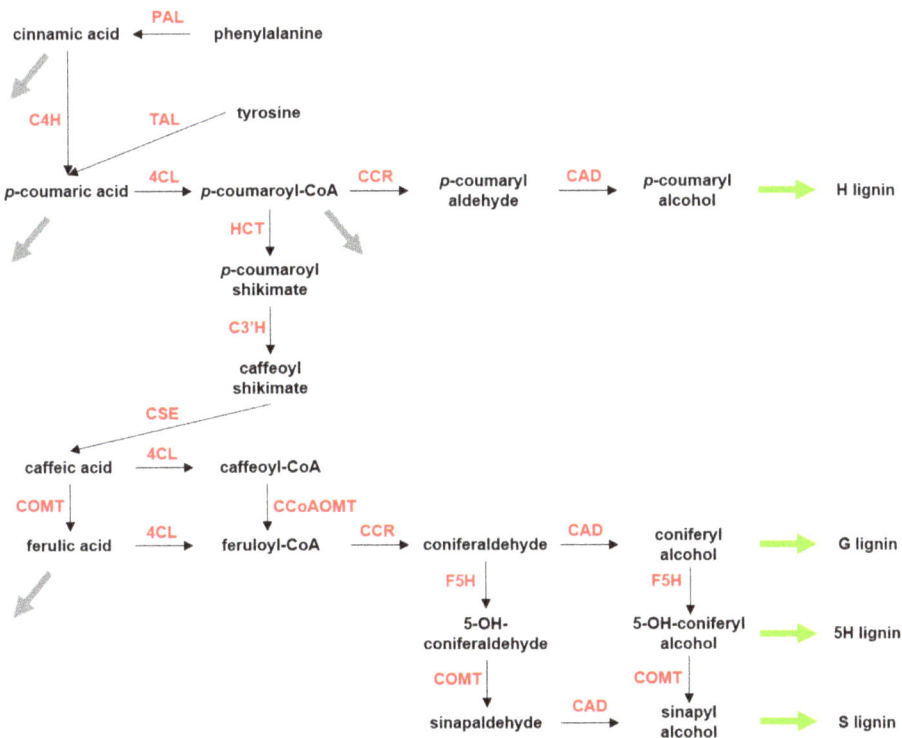

Figure 1. Lignin biosynthesis pathway in switchgrass (*Panicum virgatum*). The pathway mainly uses phenylalanine as starting compound. The reactions take place in the cytosol, and the monolignol end products are translocated to the plant cell wall, transformed into lignin monomers, and polymerized to form lignin. In switchgrass, the overall lignin composition is ~5% H, 45–50% S, and 45–50% G-lignin.

A critical first branch point is *p*-coumaroyl-CoA, where the pathway diverges toward H-lignin in one branch, and towards S- and G-lignin in the other. This divergence is important, because the normal percentage of H-lignin is quite small (~3%), and because it is apparently the ratio of S to G that significantly influences the physicochemical features of the lignin polymer. As was shown elsewhere [47], the shared enzymes between the H-lignin and S–G-lignin modules leads to crosstalk that is the result of competition between different substrates. This feature turned out to be necessary for explaining the behavior of the pathway in 4CL knockdowns. The crosstalk and other nonlinearities are indications of the difficulties of making predictions without appropriate computational tools.

In previous work [47], we developed a computational model of the lignin biosynthesis pathway in switchgrass that successfully captures the lignin profiles in wild type and four transgenic strains (4CL, COMT, CAD, and CCR knockdowns). Using experimental data on lignin content and S/G ratio in these five conditions, and designing a model within the mathematical framework of biochemical systems theory (BST), we were able to infer the structure and regulation of the lignin pathway, and to simulate its dynamics, thereby demonstrating that the model could explain all available experimental data. Furthermore, the model was validated against another transgenic strain, namely a knockdown of an inhibitor of the transcription factor PvMYB4, which had not been used to construct the model. In the following, we use this model, without changes, to predict the lignin profiles in so far untested transgenic plants, by purely computational means.

While reducing lignin content is one of the targets in bioenergy science, there is uncertainty in the literature as to whether the total lignin content plays a more important role for recalcitrance than the S/G ratio. It is not even entirely clear whether a higher or lower S/G ratio would benefit the ethanol yield. In fact, there have been contradictory reports in the literature [146,147]. The computational model is capable of simulating both scenarios, and allows optimization toward either objective.

2.4.1. A Library of Virtual Strains

The previously developed model [47] is, in fact, an ensemble of model variants that are equally capable of capturing the available experimental data in the wild type and four knockdowns, namely in COMT [43], CAD [42], CCR [148], and 4CL [40]. These data included the lignin content and composition, as well as the steady-state concentrations of several metabolites of the pathway. The data were measured in plant crude extracts, which had been collected from wild type or transgenic plants. Gene knockdowns in transgenic plants were conducted through introducing RNAi to reduce enzyme activity. The total lignin was quantified using the acetyl bromide method, and the S/G ratio was measured by thioacidolysis. Here, we use the ensemble to simulate the pathway over a wide range of perturbations, to determine the responses of the system to single or multiple increases or decreases in enzyme activities, and the consequent changes in lignin content and composition.

Single Perturbations. In the first set of computational experiments, one enzyme at a time was perturbed up to ±5-fold relative to its wild type activity. Since every scenario is simulated for the entire ensemble of models; the analysis yields many results for each scenario. Using all these results collectively, the median of total lignin content, and the median of the S/G ratio in the perturbed systems are recorded. The medians are normalized with respect to the wild type value, so that value 1 represents the wild type (see *Methods*).

The results of this analysis are shown in Figures 2 and 3. The X-axis shows the $\log_2$-fold change in the amount of a given enzyme, and the Y-axis indicates the specifically perturbed enzyme. The grouping of HCT, C3'H, and CSE represents the flux from *p*-coumaroyl-CoA to caffeic acid, as the corresponding reactions have been merged into one in our model. The color code represents the relative change in total lignin or S/G ratio, for which white depicts no change from the wild type phenotype. The green spectrum (in the panel for total lignin) and the blue spectrum (in the panel for the S/G ratio) represent reductions, while the red spectrum (in both panels) represents an increase relative to wild type. The color bar indicates the intensity of fold change in the enzymes. The greatest lignin reduction achieved is close to 50%, which is the predicted result of an 80% CAD knockdown, with 20% activity remaining. The most significant reduction and increase in S/G ratio is predicted for perturbations in F5H. Thus, the different criteria for total lignin and lignin composition point to different knockdowns.

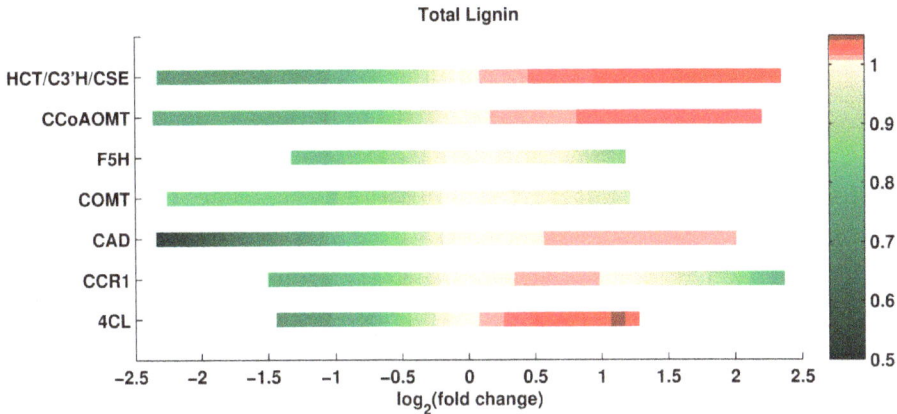

Figure 2. Total lignin in response to single enzyme perturbations. The total lignin level is color coded, where green represents decreases in total lignin, red represents increases, while white corresponds to the wild type level. CAD seems to be the most effective enzyme in reducing lignin. Note that, surprisingly, the change in lignin content is not always monotonic. Simulations show that lignin can be reduced by knocking down or overexpressing the activities of F5H, COMT and CCR.

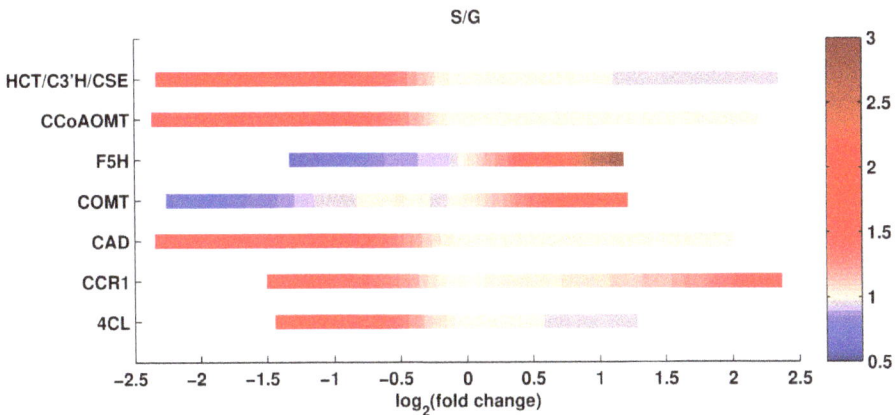

Figure 3. S/G ratios in response to single enzyme perturbations. Blue represents decreases in S/G ratios, red represents increases, while white is the wild type base level. F5H seems to be the most effective enzyme for altering the S/G ratio, both toward increases and decreases. Similar to the total lignin response, changes in S/G ratios are not necessarily monotonic: the S/G ratio increases in both knocked down and overexpressed CCR1.

Double Perturbations. It seems reasonable to surmise that simultaneous changes in two enzymes might be more effective in altering total lignin and/or the S/G ratio. Thus, we analyzed simultaneous perturbations in pairs of enzymes. The perturbations were again restricted to magnitudes of ±5-fold relative to the corresponding wild type activities. Again, every scenario was simulated with the ensemble of models, and the medians of total lignin content and of the S/G ratios were computed and normalized with respect to the wild type values. The results are shown in Figures 4 and 5. The X- and Y-axes represent the log$_2$-fold changes in the perturbed enzyme activities.

Total Lignin

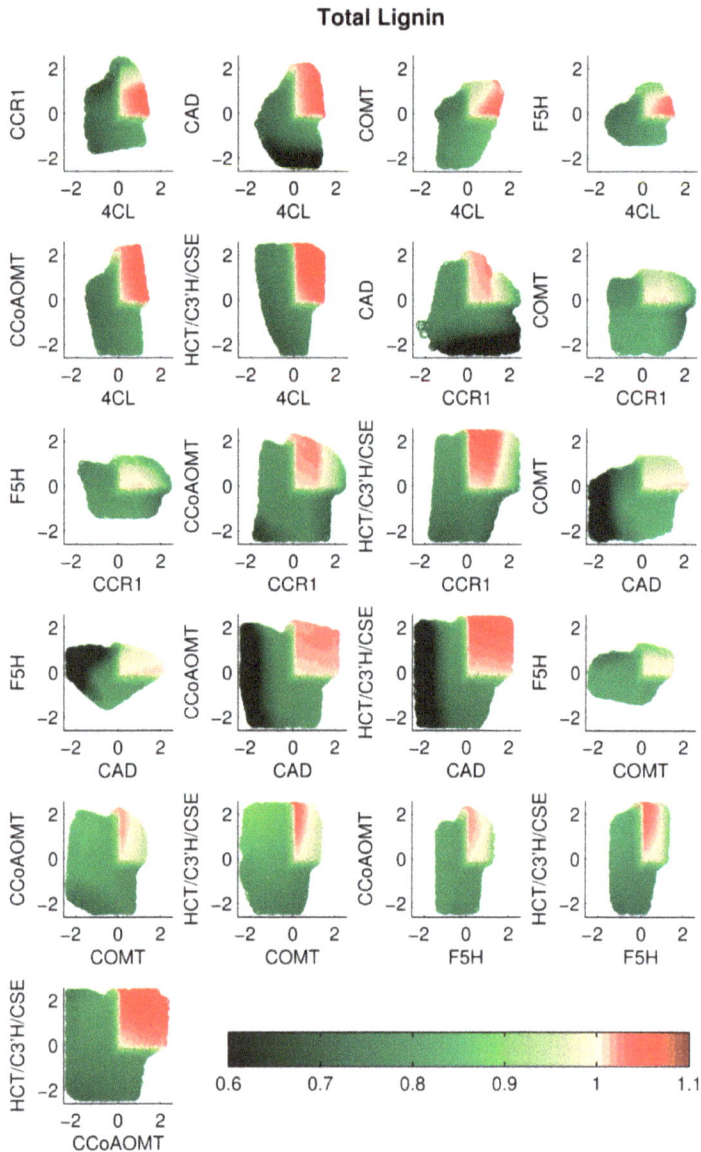

Figure 4. Total lignin in double enzyme perturbations. The color code is the same as in Figure 2. Pairs of CAD/4CL, CAD/CCR, and CAD/F5H are predicted as the most effective combinations; in particular, the pair of CAD/F5H shows strong synergism: an increase in F5H, combined with a small reduction in CAD, reduces total lignin dramatically. The nonlinear behavior of the pathway is evident in the dual-overexpression scenarios, especially in pairs, including CCR1, COMT, or F5H.

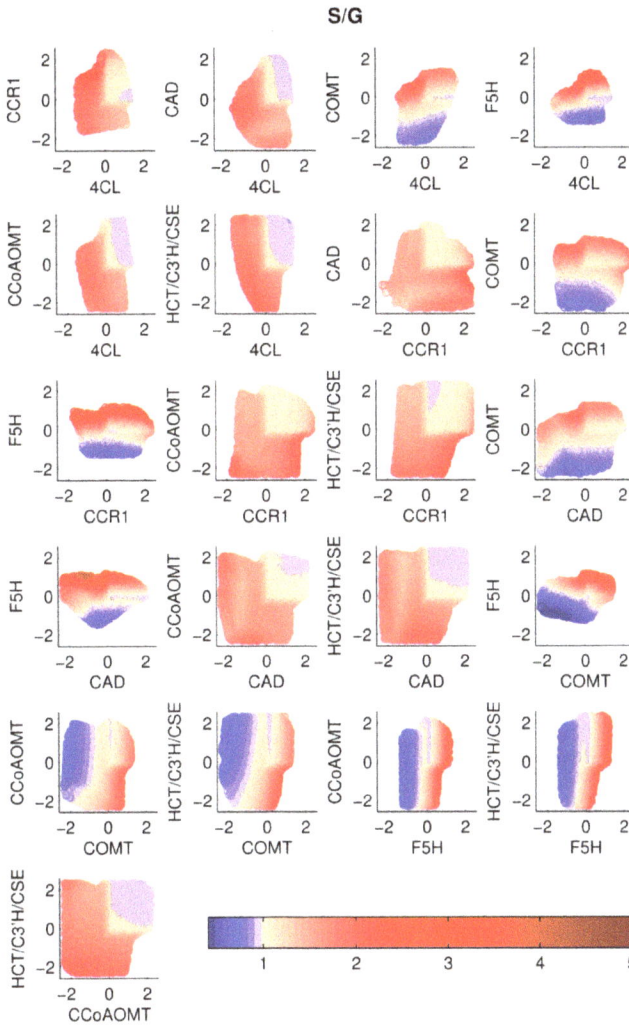

Figure 5. S/G ratios in response to two simultaneous enzyme perturbations. The color code is the same as in Figure 3. Pairs including F5H and COMT (F5H/4CL, F5H/CCR, F5H/CAD, F5H/COMT, COMT/4CL, COMT/CCR) show the highest changes in S/G ratio. In particular, F5H and COMT work well synergistically, even for moderate perturbations.

It is evident from the results that some perturbations are more effective in altering lignin content and S/G ratio, whereas the system response is more robust to others. It is also clear that many solutions reveal compromises between alterations in total lignin and the S/G ratio. If the S/G ratio is to be altered, F5H seems again to be the key enzyme, and pairs like F5H/4CL, F5H/CCR, F5H/CAD, and F5H/COMT are predicted to be most successful. At the same time, if the goal is solely to reduce lignin, irrespective of the S/G ratio, other solutions exist, including the pairs of CAD/4CL, CAD/CCR, and CAD/F5H.

Single Perturbations in a PvMYB4 Overexpression Strain. A recent study [41,149] analyzed the consequences of overexpressing the inhibitor, PvMYB4, of a transcription factor in switchgrass. The main result was an altered expression profile of many of the enzymes involved in lignin biosynthesis. Consequently, the lignin content was reduced to 40–70%, while the S/G ratio remained the same as for wild type. So far, the PvMYB4 strain has been the most effective transgenic line in reducing recalcitrance in switchgrass. To build upon this success, we combined this scenario with additional single enzyme perturbations, and investigated whether it could be possible to improve the results from the PvMYB4 transgenic strain further. As before, each enzyme was perturbed up to ±5-fold relative to the wild type level. Simulation results are shown in Figures 6 and 7. The color code is the same as for previous figures. The black square in each row represents the enzyme activity in the reference PvMYB4 perturbation. An interesting observation is that the lignin content is predicted to decrease even more than in the reference PvMYB4 experiment if CCR1 is overexpressed in this background. Additional simulations show that lignin content could be reduced further if CCoAOMT, CAD, or 4CL are reduced to even lower levels relative to the PvMYB4 background.

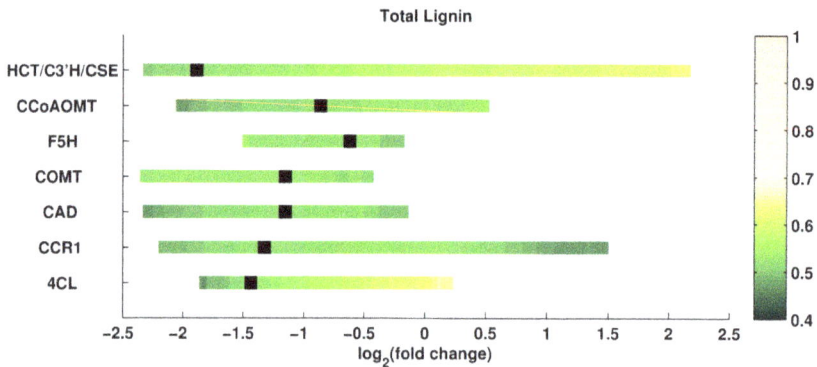

Figure 6. Total lignin in overexpressed PvMYB4 plus a single enzyme perturbation. The color code is the same as in Figure 2. The black squares represent the original amount of the corresponding enzyme in the PvMYB4 experiment. Additional overexpression of CCR is predicted to improve the total lignin results. Decreasing the level of CAD and CCoAOMT, relative to the reference PvMyb4 experiment, can reduce the total lignin further.

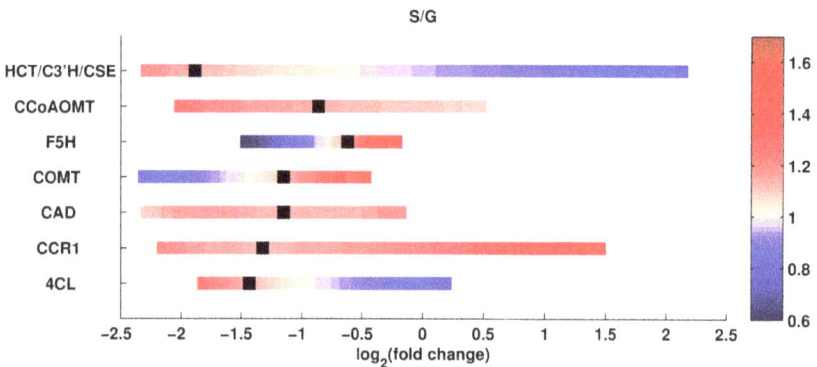

Figure 7. S/G ratio in overexpressed PvMYB4 plus an additional single enzyme perturbation. The color code is the same as in Figure 3. The black squares represent the reference amount of the corresponding enzyme in the PvMYB4 experiment. The S/G ratio can be significantly changed compared to the background PvMYB4 experiment. A change in F5H can alter the S/G ratio dramatically in a narrow perturbation interval.

2.4.2. System Optimization through Global Perturbations

So far, all perturbation profiles were determined by Monte Carlo sampling, where the goal was to find a desired combination of lignin content and composition. Now, we pursue a somewhat similar goal, except that it represents a different intent. Namely, a desired target combination of lignin content and S/G ratio is chosen a priori as the criterion for an optimization, where the goal is to determine those admissible combinatorial perturbation profiles that satisfy the criteria in an optimal manner. The main difference between this approach, and the results in Figures 2–5, is that the earlier approach tries to keep the number of enzymes to be manipulated as low as possible. Hence, the reduction in lignin content and composition is certainly not necessarily optimized. Furthermore, the changes were restricted by physiological limits, because dramatic changes in enzyme activities may lead to instability of the system, which might translate into the accumulation of toxic intermediate metabolites, or the emergence of undesired phenotypes. Therefore, the level of perturbations should be accordingly limited for at least some enzymes of the pathway.

A different approach toward identifying desirable strains is the optimization of all enzymes within physiological bounds. Although the optimized combinations might not be experimentally implementable at present, they do indicate what changes are theoretically achievable, and in which specific directions novel alterations should be pursued. Thus, this part of the project aims to compute ensembles of optimized enzyme activity patterns within physiological constraints.

As it is not clear which combination of lignin content and composition (S/G ratio) is considered optimal for a particular purpose, all enzymes in the model were simultaneously perturbed randomly up to ±5-fold, using Monte Carlo simulations. Admissible system responses were defined as stable scenarios that reached a steady state and led to an accumulation of metabolites, of at most, 6 times their normal levels, or fell at most, to 5% of the normal level. The admissible solutions were recorded and categorized based on lignin content and S/G ratio.

The results are shown in Figure 8. The three rows of subpanels represent the degree of reduction in lignin, categorized in three intervals, and the four columns of subpanels represent intervals for changes in the S/G ratio. Thus, the top row indicates the strongest reduction in lignin, and the left-most subpanel exhibits the lowest S/G ratios. Due to the randomized nature of the Monte Carlo method, we obtain many perturbation profiles that satisfy the constraints of each subpanel. All such profiles are grouped into 3 to 10 clusters, denoted by c_i. The default number of clusters is three, and each cluster contains, at most, 100 profiles. Some subpanels include more clusters, which is an indication of the abundance of admissible profiles for that range of constraints. The median of each set of profiles is computed for each cluster, and the clusters are then sorted based on the total fold change in all enzymes collectively, shown as the bottom row in each subpanel. The darker a box in the bottom row is, the more distant the strain is from wild type switchgrass. This total fold change is a measure of how distant or close a mutant strain is to the wild type. This strategy accounts for the observation that profiles closer to wild type are probably to be favored metabolically, and is in line with the concept of the minimization of metabolic adjustment [45].

Since the S/G ratio in wild type switchgrass is about 1, the central subpanels are closer to the wild type. If an experimentalist is interested in a strain with the strongest possible reduction in lignin, and the highest possible increase in the S/G ratio, the subpanel in the top right corner exhibits perturbation profiles that are predicted to achieve these criteria. Among these, cluster c_1 is the closer to wild type than c_2 and c_3 in the same subpanel. If the c_1 column is the chosen profile, the enzyme perturbation scenario is indicated by the color code. White represents the wild type, the blue spectrum represents reduced expression of the enzyme, and the red spectrum shows overexpression. The intensity of the color represents the degree of perturbation needed. The $\log_2$-fold change is indicated in the color bar.

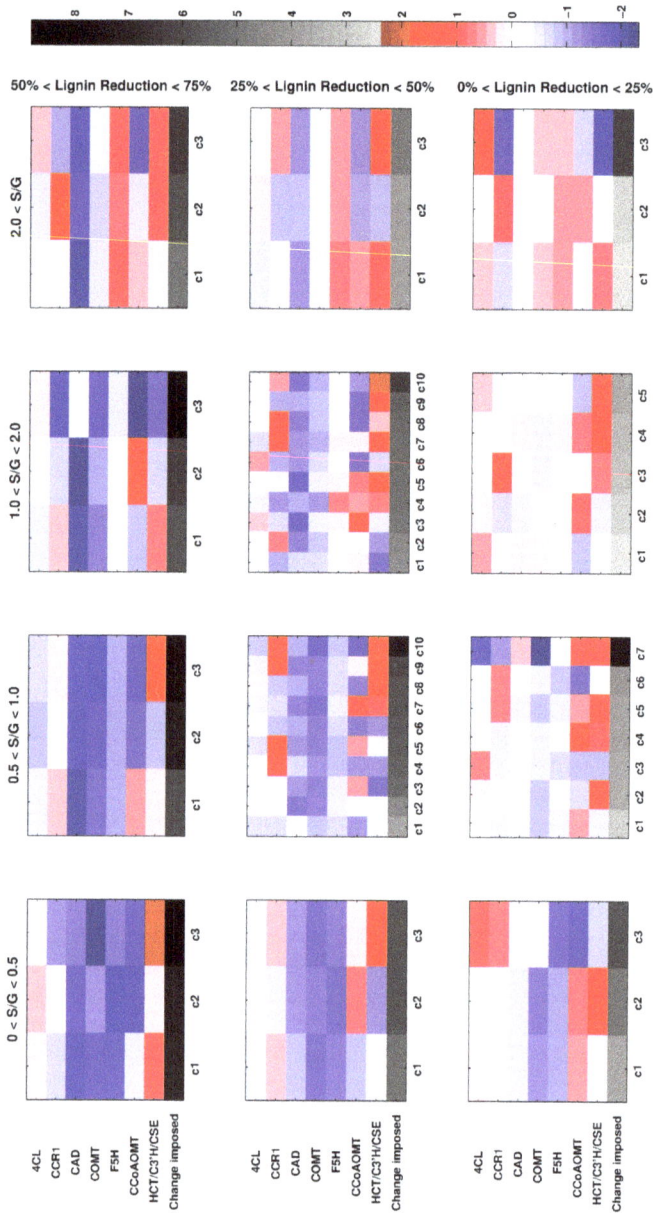

Figure 8. Global perturbation scenarios. All seven enzymes are perturbed simultaneously. The results are broken into 12 subpanels. The subpanels in the same columns share the same S/G ratio, and the subpanels in the same row share the same total lignin. Each of the subpanels includes several columns, where each column represents a cluster of perturbation vectors that are sorted based on the distance from the wild type. Cluster c1 in each subpanel is the closest perturbation scenario to the wild type. The grey scale represents the distance from the wild type, and the red/blue spectrum shows the increase/decrease in pathway enzyme. White represents the wild type, therefore, no change in the enzyme.

As a specific example, suppose that a high increase in S/G ratio and a moderate decrease in total lignin is desired, which leads us to the top right subpanel. If a medium total change in enzyme profile is allowed, we choose column c_2 as the perturbation scenario. Then, CCR1 must be overexpressed 2.8-fold, F5H must be overexpressed to 1.6-fold, the flux from the group of HCT/C3'H/CSE must be increased 1.9-fold, while 4CL, COMT, and CCoAOMT must be knocked down, as indicated in the blue range of the color scale.

Although the lignin profile in some pathway enzyme knockdowns has been measured before in vivo [40,42,148–150], our results cover seven pathway enzyme knockdowns in single, double, and combinatorial perturbations. The possibly strong ±5-fold perturbations presumably cover the realistic range of behaviors of the lignin biosynthetic pathway in response to gene knockdowns. Determining the total lignin and S/G ratio, simultaneously, provides a powerful tool for lignin researchers to choose the desired knockdown scenario based on the lignin characteristics of choice.

Figures 2–8 make it clear that the response of the system is nonlinear in some perturbation scenarios. For instance, in single enzyme perturbations (Figure 2), F5H, COMT, and CCR1 exhibit non-monotonic changes in total lignin, which means that reducing the enzyme concentration is not the only way to reduce lignin content, but that targeted overexpression may lead to the same result. In the specific case of CCR, an increase in total lignin with small degrees of enzyme overexpression, followed by a decrease in total lignin at higher levels of enzyme overexpression, is a good example of the occasional counterintuitive behavior of the pathway. The same pattern is even clearer in double knockdowns, such as the pairs of 4CL–CCR, 4CL–F5H, CCR–CAD, CCR–CCoAOMT, and COMT–CCoAOMT (Figure 3). Another interesting result is that choosing a specific perturbation scenario can retain the same amount of total lignin, while leaving room to adjust the desired S/G ratio, as it is the case for the pair 4CL–F5H: there is no substantial difference in total lignin in the left half of the figure, but there is a drastic change in S/G ratio based on the fold change in F5H. The same applies to other pairs including F5H, and also, for the combination of COMT–CCoAOMT.

The computational model turned out to be helpful for an investigation of the transgenic strain of overexpressed PvMYB4 (Figure 5). It is interesting to note that, similar to single enzyme perturbations, combinations of the profile of pathway enzymes in PvMYB4 line with overexpressed CCR can further improve the reduction in total lignin. Again, the S/G ratio is easy to manipulate, while keeping the total lignin almost unchanged; see for instance, combinations of the profile with F5H or COMT.

An added benefit of the developed library is that our results could be complemented with a record of the ethanol yield in the transgenic plants containing different total lignin contents and different S/G ratios. This record could provide the desired lignin content and S/G ratio, and with this target, one could use our results in Figures 2–8 to choose the perturbation scenario needed to achieve the target characteristics in the transgenic plant. In other words, this combination of computational results and literature information could be of value and assistance for the targeted design of transgenic plants.

While, from a technical point of view, growing transgenic plants with more than two knocked down enzymes does not seem to be practical at present, fast and inexpensive computational modeling is not really limited. Thus, it offers the opportunity to investigate more complex perturbation scenarios that could shed light onto virtually optimized transgenics. For example, one could restrict the number of enzymes to be perturbed, or permit higher levels of perturbation to achieve a more significant change in total lignin or S/G ratio (cf. [151]).

Of course, caution is advised, and it will be necessary to validate the predictions with correspondingly manipulated strains. As an example of possibly wrong predictions, a drastic change in an enzyme concentration is not a problem, theoretically, but physiological restrictions might not allow it. It could also happen that a significant change in one or two enzymes might lead to intolerable changes in fluxes, or an accumulation of toxic intermediates within, or outside, the lignin pathway. By contrast, a combination of small changes at multiple locations of the pathway is experimentally more challenging, but might avoid such issues. Thus, there is range of options for reaching the same result. Among these admissible perturbation scenarios, which give the same combination of lignin content and S/G ratio, one should presumably focus on the optimized scenario that reaches the target but deviates the least from the wild type. This overall deviation is reminiscent of the philosophy of the method of minimization of metabolic adjustment (MOMA) [45,85], which we discussed before. It may be assessed with a metric, like the Euclidian distance between the enzyme profiles in the virtual transgenic and the wild type. As one might expect, our results show that the more the lignin content is to be reduced, the further the optimized enzyme perturbation profile deviates from the wild type. Interestingly, the S/G ratio is not particularly sensitive to the distance from the wild type, and for the same lignin content, the optimized enzyme perturbation profiles for different S/G ratios are quite close to each other. In other words, it seems that altering the S/G ratio does not introduce plants with severely altered characteristics.

3. Methods

As described elsewhere [47], the lignin pathway system is modeled by a system of stoichiometric differential equations of the form

$$\frac{dX_i}{dt} = \sum_{j=1}^{k} s_{i,j} V_j \tag{1}$$

Here, each X_i is a metabolite, V_j represent fluxes, and $s_{i,j}$ are stoichiometric matrix elements. If flux V_j enters the pool of metabolite X_i, then $s_{i,j}$ is 1. If flux V_j leaves the pool of metabolite X_i, $s_{i,j}$ is -1, and if the flux V_j has no direct association with metabolite X_i, $s_{i,j}$ is zero. If all equations equal zero, the pathway operates at a steady state, where the concentrations do not change.

Each flux in Equation (1) is formulated as a general mass action (GMA) equation within the modeling framework of biochemical systems theory [114–117,152],

$$V_j = \alpha_j \prod_{r=1}^{n} X_r{}^{g_{r,j}} \prod_{r=n+1}^{n+m} X_r{}^{h_{r,j}} \tag{2}$$

where n is the number of metabolites, and m is the number of enzymes in the system. Furthermore, α_j is the *rate constant*, X_r for $1 < r < n$ is a metabolite, and X_r for $n+1 < r < n+m$ is an enzyme. The metabolites are the dependent variables of the system, whereas the enzymes are considered independent variables that do not change during a computational experiment. The parameters $g_{r,j}$ and $h_{r,j}$ are called *kinetic orders*. They determine whether a metabolite or enzyme is involved in a flux or not, and if so, in what manner and how strong. For an enzyme, it is customary to set the value of the kinetic order $h_{r,j}$ to 0 or 1, which implies that a particular enzyme is not at all involved in the process, or that the flux V_j is linearly dependent on the activity of enzyme X_r. $g_{r,j}$ can take a positive or negative value. If $g_{r,j} > 0$, the metabolite X_r is either a substrate or an activator of the flux V_j, and if $g_{r,j} < 0$, X_r is an inhibitor. Here we assume $g_{r,j}$ to range between -1 and 1, which is typical (cf. Ch. 5 of [152]). In the case here, the values of the rate constants and kinetic orders are known from the parameterized model in the previous work [47].

Due to lacking information, the steady-state concentrations of the metabolites and the enzyme activities of the pathway are not known. Hence, normalized concentrations relative to wild type concentrations are used.

3.1. A Library of Virtual Mutant Strains

As a consequence of the normalization of the variables, the X_r values for the enzymes are equal to 1 when a wild type strain is modeled. If a knockdown strain is modeled, the corresponding enzyme X_r will have a value less than 1, and if a strain with an upregulated enzyme is modeled, the corresponding enzyme X_r value will be greater than 1. Once a perturbation by down- or upregulation of an enzyme has been introduced, the system rearranges itself, and typically achieves a new steady state. At this state, at least some of the fluxes and metabolites typically assume new values. Thus, the affected flux becomes

$$V'_j = \alpha_j \prod_{r=1}^{n} X_r'^{g_{r,j}} \prod_{r=n+1}^{n+m} X'_r \qquad (3)$$

To generate a library of virtual mutants, enzyme concentrations, X_r ($n + 1 < r < n + m$), are perturbed up to ± 5-fold. If the number of enzymes to be perturbed is k, using an extensive Monte Carlo simulation, a hypercube in $\mathbb{R}^k$ is randomly sampled, and the generated set of arrays of random values is fed into the system. Since α_j and $g_{r,j}$ are already known, the differential equations can be immediately simulated for single or double alterations, or for the PvMYB4 overexpression strain combined with an additional perturbation.

3.2. Admissible Results

Each scenario corresponds to an array of perturbed enzymes. The total lignin content and S/G ratio for the scenario are recorded, if the perturbed system satisfies the following criteria:

1. The system is stable at the steady state, and reaches this state after a perturbation.
2. The steady-state values of the metabolites do not exceed a value of 6 times the wild type concentration.
3. The steady-state values of the metabolites do not fall below 5% of the wild type value.

Since an ensemble of models is used to simulate each scenario, multiple lignin profiles exist for each perturbation scenario. For visualization purposes, the median of the total lignin and the corresponding S/G ratio of the ensemble for each scenario is plotted against the perturbed enzyme(s).

3.3. Global Perturbations and Optimized Virtual Mutant Strains

All seven enzymes are perturbed simultaneously within a range of up to ± 5-fold about the wild type value. Perturbation scenarios that satisfy the criteria described in the previous section are recorded. The recorded results are arranged in a matrix based on the total lignin content and S/G ratio. The total lignin range is subdivided into three intervals, namely for 25–50%, 50–75%, and 75–100% reduction in total lignin relative to the wild type, while the S/G ratio is subdivided into four intervals, namely 0–0.5, 0.5–1, 1–2, and >2. These intervals group the results into 12 sets with different total lignin and S/G ratio characteristics.

Each square in the results matrix contains a set with numerous scenarios satisfying specific interval criteria for total lignin and S/G ratio. To facilitate the interpreting of the results, the scenarios are clustered, and the clusters are sorted based on the distance from the wild type, which reflects the overall change imposed upon the system (Figure 8). This distance is defined as the Euclidian distance between the vector of perturbed enzymes and the vector of the corresponding wild type value. The smaller the distance is, the closer is the virtual mutant is to the wild type (Figure 9).

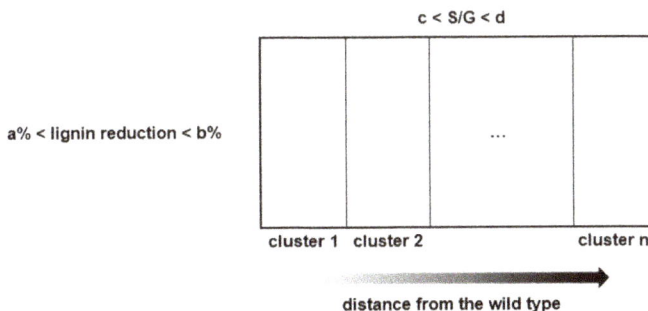

Figure 9. Clusters of results in matrix subpanels of Figure 8.

4. Discussion and Conclusions

The article consists of two parts. In the first, we described generic issues of crop modeling, and illustrated them with prominent methods and applications from the literature. Metabolic modeling in plants does not have as rich a history as in animal cells or bacteria, and many of the computational models developed so far have not been specifically tuned for the distinct physiology of plants. We addressed this issue by elaborating strategies on the basis of the most common mathematical techniques in metabolic modeling, focusing on plant applications, and discussing possible discrepancies between standard applications and the specific attributes of plants.

The hope behind this strategy of representing the field was that experimentalists and modelers may find common ground in this discussion. As plant and crop modeling matures, it seems that the current methods may have to be customized toward the genuine features of plants. This customization will have to be the outcome of an extensive, ongoing dialogue between the experimental and computational communities. Such a dialogue is not always easy, as the different communities use different languages and observe nature from different viewpoints. However, the reward of such a collaboration will be the emergence of new integrative approaches that are able to answer pressing questions of the field and guide future experimentation.

As a pertinent side issue, we discussed the reliability of mechanistic modeling using in vitro data, and the fact that researchers must take caution when drawing conclusions from such models, unless extensive validation against in vivo data supports the model performance. Computational models have been instrumental in demonstrating that even for the best known pathways, where allegedly every element of the system is known, in vitro measurements might not be representative enough to simulate the typically complex in vivo environment [103].

An intriguing aspect of computational modeling is the feature of emergence. One type of the emergence of unexpected behavior occurs when a system or organism is exposed to a new environment, in which case, a model may or may not be able to explain the observations. However, the success of computational models in predicting the emergence of some feature, although it had not been foreseen by experimental data, may sometimes be traced back to the nonlinear nature of biological systems, and the model may even offer explanations. It is clear that plant cells, due to their complexity, are susceptible to such issues.

Many mathematical techniques were reviewed in this study, and numerous other techniques exist that have not yet been applied specifically to plant metabolism. The choice of the best suited technique points to the overriding questions of any modeling effort, namely: what are the phenomena that the model is supposed to explain, and which approaches are most efficacious for analyzing the available data? The dichotomies of steady-state or dynamic, deterministic versus stochastic, mechanistic or non-mechanistic models are important choices that a modeler needs to make at the beginning of every analysis, and that ultimately define the quality of a model.

We demonstrated the potential of a model to guide future research in the second part of this article, which was dedicated to a detailed case study. Specifically, we used an earlier model to illustrate how model predictions may provide guideposts for future genomic manipulations. As we demonstrated, the computational model of the lignin biosynthetic pathway captured the complexity of the biological systems and led to insights beyond intuition. It also provided a powerful tool for simulating mutant plants with modified lignin characteristics, and prescribed single, double, combinatorial and global pathway enzyme knockdowns that were predicted to yield plant designs with specifically altered lignin content and composition. The model itself had been validated against a PvMYB4 strain before generating the libraries, but it is clear that further validation against other transgenics will improve the reliability of the presented results, and may improve the model itself.

Acknowledgments: This work was supported by DOE-BESC grant DE-AC05-00OR22725 (PI: Paul Gilna). BESC, the BioEnergy Science Center, is a U.S. Department of Energy Bioenergy Research Center supported by the Office of Biological and Environmental Research in the DOE Office of Science.

Author Contributions: M.F. designed the model as well as the computational experiments. E.O.V. supervised the project and contributed to all computational aspects. Both authors collaborated on the writing of the article.

Conflicts of Interest: The authors declare no conflicts of interest.

Abbreviations

PAL	L-phenylalanine ammonia-lyase
C4H	cinnamate 4-hydroxylase
4CL	4-coumarate:CoA ligase
CCR1	cinnamoyl CoA reductase 1
CAD	cinnamyl alcohol dehydrogenase
HCT	hydroxycinnamoyl-CoA:shikimate hydroxycinnamoyl transferase
C3′H	p-coumaroyl shikimate 3′-hydroxylase
CSE	caffeoyl shikimate esterase
COMT	caffeic acid O-methyltransferase
CCoAOMT	caffeoyl CoA O-methyltransferase
F5H	ferulate 5-hydroxylase
ER	endoplasmic reticulum
BST	biochemical systems theory
GMA	generalized mass action
FBA	flux balance analysis

References

1. Doebley, J. Teosinte As a GraHin Crop. Available online: http://teosinte.wisc.edu/grain_Crop.html (accessed on 5 August 2017).
2. List of Sequenced Plant Genomes. Available online: http://en.wikipedia.org/wiki/List_of_sequenced_plant_genomes#Gymnosperm (accessed on 5 August 2017).
3. The Human Genome Project Completion. Available online: http://www.genome.gov/11006943/human-genome-project-completion-frequently-asked-questions/ (accessed on 5 August 2017).
4. Williams, T.C.R.; Miguet, L.; Masakapalli, S.K.; Kruger, N.J.; Sweetlove, L.J.; Ratcliffe, R.G. Metabolic network fluxes in heterotrophic arabidopsis cells: Stability of the flux distribution under different oxygenation conditions. *Plant Physiol.* **2008**, *148*, 704–718. [CrossRef] [PubMed]
5. Yuan, J.S.; Galbraith, D.W.; Dai, S.Y.; Griffin, P.; Stewart, C.N. Plant systems biology comes of age. *Trends Plant Sci.* **2008**, *13*, 165–171. [CrossRef] [PubMed]
6. Human Metabolome Database. Available online: http://www.hmdb.ca/statistics (accessed on 5 August 2017).
7. Bionumbers. Available online: http://bionumbers.hms.harvard.edu/bionumber.aspx?id=105634&ver=4 (accessed on 5 August 2017).
8. Dixon, R.A.; Strack, D. Phytochemistry meets genome analysis, and beyond. *Phytochemistry* **2003**, *62*, 815–816. [CrossRef]
9. Saito, K.; Matsuda, F. Metabolomics for functional genomics, systems biology, and biotechnology. *Annu. Rev. Plant Biol.* **2010**, *61*, 463–489. [CrossRef] [PubMed]
10. Cao, H.X.; Wang, W.; Le, H.T.T.; Vu, G.T.H. The power of CRISPR-Cas9-induced genome editing to speed up plant breeding. *Int. J. Genom.* **2016**, *2016*, 10. [CrossRef] [PubMed]
11. Shan, Q.; Wang, Y.; Li, J.; Zhang, Y.; Chen, K.; Liang, Z.; Zhang, K.; Liu, J.; Xi, J.J.; Qiu, J.-L.; et al. Targeted genome modification of crop plants using a CRISPR-Cas system. *Nat. Biotech.* **2013**, *31*, 686–688. [CrossRef] [PubMed]
12. Nekrasov, V.; Staskawicz, B.; Weigel, D.; Jones, J.D.G.; Kamoun, S. Targeted mutagenesis in the model plant nicotiana benthamiana using Cas9 rna-guided endonuclease. *Nat. Biotech.* **2013**, *31*, 691–693. [CrossRef] [PubMed]
13. Li, J.F.; Norville, J.E.; Aach, J.; McCormack, M.; Zhang, D.; Bush, J.; Church, G.M.; Sheen, J. Multiplex and homologous recombination-mediated genome editing in *Arabidopsis* and *Nicotiana benthamiana* using guide rna and Cas9. *Nat. Biotechnol.* **2013**, *31*, 688–691. [CrossRef] [PubMed]
14. Cai, Y.; Chen, L.; Liu, X.; Guo, C.; Sun, S.; Wu, C.; Jiang, B.; Han, T.; Hou, W. CRISPR/Cas9-mediated targeted mutagenesis of gmft2a delays flowering time in soya bean. *Plant Biotechnol. J.* **2017**. [CrossRef] [PubMed]
15. Tian, S.; Jiang, L.; Gao, Q.; Zhang, J.; Zong, M.; Zhang, H.; Ren, Y.; Guo, S.; Gong, G.; Liu, F.; et al. Efficient CRISPR-Cas9-based gene knockout in watermelon. *Plant Cell Rep.* **2017**, *36*, 399–406. [CrossRef] [PubMed]

16. Soyk, S.; Muller, N.A.; Park, S.J.; Schmalenbach, I.; Jiang, K.; Hayama, R.; Zhang, L.; Van Eck, J.; Jimenez-Gomez, J.M. Variation in the flowering gene self pruning 5g promotes day-neutrality and early yield in tomato. *Nat. Genet.* **2017**, *49*, 162–168. [CrossRef] [PubMed]

17. Aharoni, A.; Galili, G. Metabolic engineering of the plant primary-secondary metabolism interface. *Curr. Opin. Biotechnol.* **2011**, *22*, 239–244. [CrossRef] [PubMed]

18. Ratcliffe, R.G.; Shachar-Hill, Y. Measuring multiple fluxes through plant metabolic networks. *Plant J. Cell Mol. Biol.* **2006**, *45*, 490–511. [CrossRef] [PubMed]

19. Morgan, J.A.; Rhodes, D. Mathematical modeling of plant metabolic pathways. *Metab. Eng.* **2002**, *4*, 80–89. [CrossRef] [PubMed]

20. Sweetlove, L.J.; Williams, T.C.; Cheung, C.Y.; Ratcliffe, R.G. Modelling metabolic co(2) evolution—A fresh perspective on respiration. *Plant Cell Environ.* **2013**, *36*, 1631–1640. [CrossRef] [PubMed]

21. Nepali, M.R. Polyploidy Breeding. Available online: http://mukeshramjalipb.blogspot.com/2013/03/polyploidy-breeding.html (accessed on 5 August 2017).

22. Meru, G. Polyploidy. Available online: http://plantbreeding.coe.uga.edu/index.php?title=5._Polyploidy (accessed on 5 August 2017).

23. Lukhtanov, V.A. The blue butterfly *Polyommatus* (*plebicula*) *atlanticus* (*lepidoptera, lycaenidae*) holds the record of the highest number of chromosomes in the non-polyploid eukaryotic organisms. *Comp. Cytogenet.* **2015**, *9*, 683–690. [CrossRef] [PubMed]

24. Janick, J.; American Society for Horticultural Science. *Plant Breeding Reviews*; Wiley Blackwell: Hoboken, NJ, USA, 2009; Volume 31.

25. Yu, J.; Wang, J.; Lin, W.; Li, S.; Li, H.; Zhou, J.; Ni, P.; Dong, W.; Hu, S.; Zeng, C.; et al. The genomes of *Oryza sativa*: A history of duplications. *PLoS Biol.* **2005**, *3*, e38. [CrossRef] [PubMed]

26. Arnold, A.; Nikoloski, Z. In search for an accurate model of the photosynthetic carbon metabolism. *Math. Comput. Simul.* **2014**, *96*, 171–194. [CrossRef]

27. Szecowka, M.; Heise, R.; Tohge, T.; Nunes-Nesi, A.; Vosloh, D.; Huege, J.; Feil, R.; Lunn, J.; Nikoloski, Z.; Stitt, M.; et al. Metabolic fluxes in an illuminated *Arabidopsis* rosette. *Plant Cell* **2013**, *25*, 694–714. [CrossRef] [PubMed]

28. Zhu, X.G.; Wang, Y.; Ort, D.R.; Long, S.P. E-photosynthesis: A comprehensive dynamic mechanistic model of c3 photosynthesis: From light capture to sucrose synthesis. *Plant Cell Environ.* **2013**, *36*, 1711–1727. [CrossRef] [PubMed]

29. Arnold, A.; Nikoloski, Z. A quantitative comparison of calvin-benson cycle models. *Trends Plant Sci.* **2011**, *16*, 676–683. [CrossRef] [PubMed]

30. Cheung, C.Y.M.; Poolman, M.G.; Fell, D.A.; Ratcliffe, R.G.; Sweetlove, L.J. A diel flux balance model captures interactions between light and dark metabolism during day-night cycles in c-3 and crassulacean acid metabolism leaves. *Plant Physiol.* **2014**, *165*, 917–929. [CrossRef] [PubMed]

31. Boyle, N.R.; Morgan, J.A. Computation of metabolic fluxes and efficiencies for biological carbon dioxide fixation. *Metab. Eng.* **2011**, *13*, 150–158. [CrossRef] [PubMed]

32. Guo, Y.; Tan, J.L. A kinetic model structure for delayed fluorescence from plants. *Biosystems* **2009**, *95*, 98–103. [CrossRef] [PubMed]

33. Pearcy, R.W.; Gross, L.J.; He, D. An improved dynamic model of photosynthesis for estimation of carbon gain in sunfleck light regimes. *Plant Cell Environ.* **1997**, *20*, 411–424. [CrossRef]

34. Poolman, M.G.; Miguet, L.; Sweetlove, L.J.; Fell, D.A. A genome-scale metabolic model of *Arabidopsis* and some of its properties. *Plant Physiol.* **2009**, *151*, 1570–1581. [CrossRef] [PubMed]

35. Lakshmanan, M.; Zhang, Z.Y.; Mohanty, B.; Kwon, J.Y.; Choi, H.Y.; Nam, H.J.; Kim, D.I.; Lee, D.Y. Elucidating rice cell metabolism under flooding and drought stresses using flux-based modeling and analysis. *Plant Physiol.* **2013**, *162*, 2140–2150. [CrossRef] [PubMed]

36. Sweetlove, L.J.; Beard, K.F.M.; Nunes-Nesi, A.; Fernie, A.R.; Ratcliffe, R.G. Not just a circle: Flux modes in the plant tca cycle. *Trends Plant Sci.* **2010**, *15*, 462–470. [CrossRef] [PubMed]

37. Baghalian, K.; Hajirezaei, M.R.; Schreiber, F. Plant metabolic modeling: Achieving new insight into metabolism and metabolic engineering. *Plant Cell* **2014**, *26*, 3847–3866. [CrossRef] [PubMed]

38. Rohwer, J.M. Kinetic modelling of plant metabolic pathways. *J. Exp. Bot.* **2012**, *63*, 2275–2292. [CrossRef] [PubMed]

39. Boerjan, W.; Ralph, J.; Baucher, M. Lignin biosynthesis. *Annu. Rev. Plant Biol.* **2003**, *54*, 519–546. [CrossRef] [PubMed]

40. Xu, B.; Escamilla-Treviño, L.L.; Sathitsuksanoh, N.; Shen, Z.; Shen, H.; Zhang, Y.H.; Dixon, R.A.; Zhao, B. Silencing of 4-coumarate: Coenzyme a ligase in switchgrass leads to reduced lignin content and improved fermentable sugar yields for biofuel production. *New Phytol.* **2011**, *192*, 611–625. [CrossRef] [PubMed]

41. Shen, H.; Poovaiah, C.R.; Ziebell, A.; Tschaplinski, T.J.; Pattathil, S.; Gjersing, E.; Engle, N.L.; Katahira, R.; Pu, Y.; Sykes, R.; et al. Enhanced characteristics of genetically modified switchgrass (*Panicum virgatum* L.) for high biofuel production. *Biotechnol. Biofuels* **2013**, *6*, 71. [CrossRef] [PubMed]

42. Fu, C.X.; Xiao, X.R.; Xi, Y.J.; Ge, Y.X.; Chen, F.; Bouton, J.; Dixon, R.A.; Wang, Z.Y. Downregulation of cinnamyl alcohol dehydrogenase (cad) leads to improved saccharification efficiency in switchgrass. *Bioenerg. Res.* **2011**, *4*, 153–164. [CrossRef]

43. Tschaplinski, T.J.; Standaert, R.F.; Engle, N.L.; Martin, M.Z.; Sangha, A.K.; Parks, J.M.; Smith, J.C.; Samuel, R.; Jiang, N.; Pu, Y.; et al. Down-regulation of the caffeic acid o-methyltransferase gene in switchgrass reveals a novel monolignol analog. *Biotechnol. Biofuels* **2012**, *5*, 71. [CrossRef] [PubMed]

44. Lee, Y.; Voit, E.O. Mathematical modeling of monolignol biosynthesis in populus xylem. *Math. Biosci.* **2010**, *228*, 78–89. [CrossRef] [PubMed]

45. Lee, Y.; Chen, F.; Gallego-Giraldo, L.; Dixon, R.A.; Voit, E.O. Integrative analysis of transgenic alfalfa (*Medicago sativa* L.) suggests new metabolic control mechanisms for monolignol biosynthesis. *PLoS Comput. Biol.* **2011**, *7*, e1002047. [CrossRef] [PubMed]

46. Wang, J.P.; Naik, P.P.; Chen, H.C.; Shi, R.; Lin, C.Y.; Liu, J.; Shuford, C.M.; Li, Q.; Sun, Y.H.; Tunlaya-Anukit, S.; et al. Complete proteomic-based enzyme reaction and inhibition kinetics reveal how monolignol biosynthetic enzyme families affect metabolic flux and lignin in populus trichocarpa. *Plant Cell* **2014**, *26*, 894–914. [CrossRef] [PubMed]

47. Faraji, M.; Fonseca, L.L.; Escamilla-Treviño, L.; Dixon, R.A.; Voit, E.O. Computational inference of the structure and regulation of the lignin pathway in panicum virgatum. *Biotechnol. Biofuels* **2015**, *8*, 151. [CrossRef] [PubMed]

48. Amthor, J.S. Efficiency of lignin biosynthesis: A quantitative analysis. *Ann. Bot.* **2003**, *91*, 673–695. [CrossRef] [PubMed]

49. Saha, R.; Suthers, P.F.; Maranas, C.D. Zea mays irs1563: A comprehensive genome-scale metabolic reconstruction of maize metabolism. *PLoS ONE* **2011**, *6*, e21784. [CrossRef] [PubMed]

50. Faraji, M.; Fonseca, L.L.; Escamilla-Trevino, L.; Barros-Rios, J.; Engle, N.; Yang, Z.K.; Tschaplinski, T.J.; Dixon, R.A.; Voit, E.O. Mathematical models of lignin biosynthesis. *Biotechnol. Biofuels* **2017**. under review.

51. Marshall-Colon, A.; Long, S.P.; Allen, D.K.; Allen, G.; Beard, D.A.; Benes, B.; von Caemmerer, S.; Christensen, A.J.; Cox, D.J.; Hart, J.C.; et al. Crops in silico: Generating virtual crops using an integrative and multi-scale modeling platform. *Front. Plant Sci.* **2017**, *8*. [CrossRef] [PubMed]

52. Crops in Silico. Available online: http://cropsinsilico.org/uiucncsa/ (accessed on 5 August 2017).

53. Bogart, E.; Myers, C.R. Multiscale metabolic modeling of c4 plants: Connecting nonlinear genome-scale models to leaf-scale metabolism in developing maize leaves. *PLoS ONE* **2016**, *11*, e0151722. [CrossRef] [PubMed]

54. WIMOVAC (Windows Intuitive Model of Vegetation Response to Atmospheric and Climate Change). Available online: http://www.life.illinois.edu/plantbio/wimovac/ (accessed on 5 August 2017).

55. SOYSIM—Soybean Growth Simulation Model. Available online: http://soysim.unl.edu/ (accessed on 5 August 2017).

56. Voit, E.O. Models-of-data and models-of-processes in the post-genomic era. *Math. Biosci.* **2002**, *180*, 263–274. [CrossRef]

57. Wiechert, W. C-13 metabolic flux analysis. *Metab. Eng.* **2001**, *3*, 195–206. [CrossRef] [PubMed]

58. Wiechert, W.; Mollney, M.; Petersen, S.; de Graaf, A.A. A universal framework for c-13 metabolic flux analysis. *Metab. Eng.* **2001**, *3*, 265–283. [CrossRef] [PubMed]

59. Maarleveld, T.R.; Khandelwal, R.A.; Olivier, B.G.; Teusink, B.; Bruggeman, F.J. Basic concepts and principles of stoichiometric modeling of metabolic networks. *Biotechnol. J.* **2013**, *8*, 997–1008. [CrossRef] [PubMed]

60. Libourel, I.G.; Shachar-Hill, Y. Metabolic flux analysis in plants: From intelligent design to rational engineering. *Ann. Rev. Plant Biol.* **2008**, *59*, 625–650. [CrossRef] [PubMed]

61. Kruger, N.J.; Ratcliffe, R.G. Insights into plant metabolic networks from steady-state metabolic flux analysis. *Biochimie* **2009**, *91*, 697–702. [CrossRef] [PubMed]

62. Allen, D.K.; Libourel, I.G.; Shachar-Hill, Y. Metabolic flux analysis in plants: Coping with complexity. *Plant Cell Environ.* **2009**, *32*, 1241–1257. [CrossRef] [PubMed]

63. Schwender, J.; Goffman, F.; Ohlrogge, J.B.; Shachar-Hill, Y. Rubisco without the calvin cycle improves the carbon efficiency of developing green seeds. *Nature* **2004**, *432*, 779–782. [CrossRef] [PubMed]

64. Sweetlove, L.J.; Ratcliffe, R.G. Flux-balance modeling of plant metabolism. *Front. Plant Sci.* **2011**, *2*. [CrossRef] [PubMed]

65. Varma, A.; Palsson, B.O. Metabolic flux balancing—Basic concepts, scientific and practical use. *Bio-Technology* **1994**, *12*, 994–998. [CrossRef]

66. Edwards, J.S.; Palsson, B.O. Systems properties of the *Haemophilus influenzae* Rd metabolic genotype. *J. Biol. Chem.* **1999**, *274*, 17410–17416. [CrossRef] [PubMed]

67. Heinrich, R.; Schuster, S. *The Regulation of Cellular Systems*; Chapman & Hall: New York, NY, USA, 1996; vix, 327p.

68. Gavalas, G.R. *Nonlinear Differential Equations of Chemically Reacting Systems*; Springer Verlag: New York, NY, USA, 1968; 106p.

69. Palsson, B. *Systems Biology: Properties of Reconstructed Networks*; Cambridge University Press: New York, NY, USA, 2006; xii, 322p.

70. Mahadevan, R.; Schilling, C.H. The effects of alternate optimal solutions in constraint-based genome-scale metabolic models. *Metab. Eng.* **2003**, *5*, 264–276. [CrossRef] [PubMed]

71. Schuster, S.H.S. On elementary flux modes in biochemical reaction systems at steady state. *J. Biol. Syst.* **1994**, *2*, 165–182. [CrossRef]

72. Trinh, C.T.; Wlaschin, A.; Srienc, F. Elementary mode analysis: A useful metabolic pathway analysis tool for characterizing cellular metabolism. *Appl. Microbiol. Biotechnol.* **2009**, *81*, 813–826. [CrossRef] [PubMed]

73. Kruger, N.J.; Masakapalli, S.K.; Ratcliffe, R.G. Strategies for investigating the plant metabolic network with steady-state metabolic flux analysis: Lessons from an *Arabidopsis* cell culture and other systems. *J. Exp. Bot.* **2012**, *63*, 2309–2323. [CrossRef] [PubMed]

74. Kacser, H.; Burns, J.A. The control of flux. *Symp. Soc. Exp. Biol.* **1973**, *27*, 65–104. [CrossRef] [PubMed]

75. Heinrich, R.; Rapoport, T.A. A linear steady-state treatment of enzymatic chains. Critique of the crossover theorem and a general procedure to identify interaction sites with an effector. *Eur. J. Biochem.* **1974**, *42*, 97–105. [CrossRef] [PubMed]

76. Heinrich, R.; Rapoport, T.A. A linear steady-state treatment of enzymatic chains. General properties, control and effector strength. *Eur. J. Biochem.* **1974**, *42*, 89–95. [CrossRef] [PubMed]

77. Fell, D.A. Metabolic control analysis: A survey of its theoretical and experimental development. *Biochem. J.* **1992**, *286 Pt 2*, 313–330. [CrossRef] [PubMed]

78. Orth, J.D.; Thiele, I.; Palsson, B.O. What is flux balance analysis? *Nat. Biotechnol.* **2010**, *28*, 245–248. [CrossRef] [PubMed]

79. Páez Melo, D.O.; Jay-Pang Moncada, R.; Winck, F.V.; Barrios, A.F.G. In silico analysis for biomass synthesis under different CO_2 levels for *Chlamydomonas reinhardtii* utilizing a flux balance analysis approach. In *Advances in Intelligent Systems and Computing*; Pietka, E., Ed.; Springer International Publishing: Cham, Switzerland, 2014; Volume 232, pp. 279–285.

80. Chang, R.L.; Ghamsari, L.; Manichaikul, A.; Hom, E.F.; Balaji, S.; Fu, W.; Shen, Y.; Hao, T.; Palsson, B.O.; Salehi-Ashtiani, K.; et al. Metabolic network reconstruction of *Chlamydomonas* offers insight into light-driven algal metabolism. *Mol. Syst. Biol.* **2011**, *7*, 518. [CrossRef] [PubMed]

81. Flassig, R.J.; Fachet, M.; Hoffner, K.; Barton, P.I.; Sundmacher, K. Dynamic flux balance modeling to increase the production of high-value compounds in green microalgae. *Biotechnol. Biofuels* **2016**, *9*, 165. [CrossRef] [PubMed]

82. Sengupta, T.; Bhushan, M.; Wangikar, P.P. Metabolic modeling for multi-objective optimization of ethanol production in a *Synechocystis* mutant. *Photosynth. Res.* **2013**, *118*, 155–165. [CrossRef] [PubMed]

83. Villaverde, A.F.; Bongard, S.; Mauch, K.; Balsa-Canto, E.; Banga, J.R. Metabolic engineering with multi-objective optimization of kinetic models. *J. Biotechnol.* **2016**, *222*, 1–8. [CrossRef] [PubMed]

84. Barros, J.; Serrani-Yarce, J.C.; Chen, F.; Baxter, D.; Venables, B.J.; Dixon, R.A. Role of bifunctional ammonia-lyase in grass cell wall biosynthesis. *Nat. Plants* **2016**, *2*, 16050. [CrossRef] [PubMed]

85. Segre, D.; Vitkup, D.; Church, G.M. Analysis of optimality in natural and perturbed metabolic networks. *Proc. Natl. Acad. Sci. USA* **2002**, *99*, 15112–15117. [CrossRef] [PubMed]

86. Hay, J.; Schwender, J. Metabolic network reconstruction and flux variability analysis of storage synthesis in developing oilseed rape (*Brassica napus* L.) embryos. *Plant J. Cell Mol. Biol.* **2011**, *67*, 526–541. [CrossRef] [PubMed]

87. Hay, J.; Schwender, J. Computational analysis of storage synthesis in developing *Brassica napus* L. (oilseed rape) embryos: Flux variability analysis in relation to (1)(3)c metabolic flux analysis. *Plant J. Cell Mol. Biol.* **2011**, *67*, 513–525. [CrossRef] [PubMed]

88. Steuer, R.; Nesi, A.N.; Fernie, A.R.; Gross, T.; Blasius, B.; Selbig, J. From structure to dynamics of metabolic pathways: Application to the plant mitochondrial tca cycle. *Bioinformatics* **2007**, *23*, 1378–1385. [CrossRef] [PubMed]

89. Schuster, S.; Dandekar, T.; Fell, D.A. Detection of elementary flux modes in biochemical networks: A promising tool for pathway analysis and metabolic engineering. *Trends Biotechnol.* **1999**, *17*, 53–60. [CrossRef]

90. Llaneras, F.; Pico, J. Which metabolic pathways generate and characterize the flux space? A comparison among elementary modes, extreme pathways and minimal generators. *J. Biomed. Biotechnol.* **2010**, *2010*, 753904. [CrossRef] [PubMed]

91. Sherry, A.D.; Malloy, C.R. Integration of 13c isotopomer methods and hyperpolarization provides a comprehensive picture of metabolism. In *eMagRes*; John Wiley & Sons, Ltd.: Chichester, UK, 2007.

92. Roscher, A.; Kruger, N.J.; Ratcliffe, R.G. Strategies for metabolic flux analysis in plants using isotope labelling. *J. Biotechnol.* **2000**, *77*, 81–102. [CrossRef]

93. Dieuaide-Noubhani, M.; Alonso, A.P. Application of metabolic flux analysis to plants. *Methods Mol. Biol.* **2014**, *1090*, 1–17. [PubMed]

94. Moreno-Sanchez, R.; Saavedra, E.; Rodriguez-Enriquez, S.; Olin-Sandoval, V. Metabolic control analysis: A tool for designing strategies to manipulate metabolic pathways. *J. Biomed. Biotechnol.* **2008**, *2008*, 597913. [CrossRef] [PubMed]

95. Ap Rees, T.; Hill, S.A. Metabolic control analysis of plant metabolism. *Plant Cell Environ.* **1994**, *17*, 587–599. [CrossRef]

96. Giersch, C.; Lämmel, D.; Farquhar, G. Control analysis of photosynthetic CO_2 fixation. *Photosynth. Res.* **1990**, *24*, 151–165. [PubMed]

97. Waage, P.; Gulberg, C.M. Studies concerning affinity. *J. Chem. Educ.* **1986**, *63*, 1044. [CrossRef]

98. Farré, G.; Maiam Rivera, S.; Alves, R.; Vilaprinyo, E.; Sorribas, A.; Canela, R.; Naqvi, S.; Sandmann, G.; Capell, T.; Zhu, C.; et al. Targeted transcriptomic and metabolic profiling reveals temporal bottlenecks in the maize carotenoid pathway that may be addressed by multigene engineering. *Plant J.* **2013**, *75*, 441–455. [CrossRef] [PubMed]

99. Bai, C.; Rivera, S.M.; Medina, V.; Alves, R.; Vilaprinyo, E.; Sorribas, A.; Canela, R.; Capell, T.; Sandmann, G.; Christou, P.; et al. An in vitro system for the rapid functional characterization of genes involved in carotenoid biosynthesis and accumulation. *Plant J.* **2014**, *77*, 464–475. [CrossRef] [PubMed]

100. Michaelis, L.; Menten, M.L. Die Kinetik der Invertinwirkung. *Biochem. Z.* **1913**, *49*, 333–369.

101. Henri, V. *Lois Générales de L'action des Diastases*; Librairie Scientifique A. Hermann: Paris, France, 1903; viii, 129p.

102. Cornish-Bowden, A. One hundred years of Michaelis–Menten kinetics. *Perspect. Sci.* **2015**, *4*, 3–9. [CrossRef]

103. Teusink, B.; Passarge, J.; Reijenga, C.A.; Esgalhado, E.; van der Weijden, C.C.; Schepper, M.; Walsh, M.C.; Bakker, B.M.; van Dam, K.; Westerhoff, H.V.; et al. Can yeast glycolysis be understood in terms of in vitro kinetics of the constituent enzymes? Testing biochemistry. *Eur. J. Biochem.* **2000**, *267*, 5313–5329. [CrossRef] [PubMed]

104. Schulz, A.R. *Enzyme Kinetics: From Diastase to Multi-Enzyme Systems*; Cambridge University Press: New York, NY, USA, 1994; x, 246p.

105. Hill, A.V. The possible effects of the aggregation of the molecules of haemoglobin on its dissociation curves. *J. Physiol.* **1910**, *40*, 4–7.

106. Nag, A.; Lunacek, M.; Graf, P.A.; Chang, C.H. Kinetic modeling and exploratory numerical simulation of chloroplastic starch degradation. *BMC Syst. Biol.* **2011**, *5*, 94. [CrossRef] [PubMed]

107. Cornish-Bowden, A. *Fundamentals of Enzyme Kinetics*, 3rd ed.; Portland Press: London, UK, 2004; xvi, 422p.

108. Voit, E.O. The best models of metabolism. *Wiley Interdisciplin. Rev. Syst. Biol. Med.* **2017**. [CrossRef] [PubMed]

109. Lee, Y.; Escamilla-Trevino, L.; Dixon, R.A.; Voit, E.O. Functional analysis of metabolic channeling and regulation in lignin biosynthesis: A computational approach. *PLoS Comput. Biol.* **2012**, *8*, e1002769. [CrossRef] [PubMed]

110. Steuer, R.; Gross, T.; Selbig, J.; Blasius, B. Structural kinetic modeling of metabolic networks. *Proc. Natl. Acad. Sci. USA* **2006**, *103*, 11868–11873. [CrossRef] [PubMed]

111. Goel, G.; Chou, I.C.; Voit, E.O. System estimation from metabolic time-series data. *Bioinformatics* **2008**, *24*, 2505–2511. [CrossRef] [PubMed]

112. Faraji, M.; Voit, E.O. Nonparametric dynamic modeling. *Math. Biosci.* **2017**, *287*, 130–146. [CrossRef] [PubMed]

113. Faraji, M.; Voit, E.O. Stepwise inference of likely dynamic flux distributions from metabolic time series data. *Bioinformatics* **2017**. [CrossRef] [PubMed]

114. Savageau, M.A. Biochemical systems analysis. I. Some mathematical properties of the rate law for the component enzymatic reactions. *J. Theor. Biol.* **1969**, *25*, 365–369. [CrossRef]

115. Savageau, M.A. Biochemical systems analysis. Ii. The steady-state solutions for an n-pool system using a power-law approximation. *J. Theor. Biol.* **1969**, *25*, 370–379. [CrossRef]

116. Savageau, M.A. *Biochemical Systems Analysis: A Study of Function and Design in Molecular Biology*; Addison-Wesley Pub. Co. Advanced Book Program: Reading, Massachusetts, USA, 1976; xvii, 379p.

117. Voit, E.O. Biochemical systems theory: A review. *ISRN Biomath.* **2013**, *2013*, 53. [CrossRef]

118. Voit, E.O. Dynamics of self-thinning plant stands. *Ann. Bot.* **1988**, *62*, 67–78. [CrossRef]

119. Torsella, J.; Bin Razali, A. An analysis of forestry data. In *Canonical Nonlinear Modeling: S-System Approach to Understanding Complexity*; Voit, E.O., Ed.; Van Nostrand Reinhold: New York, NY, USA, 1991; pp. 181–199.

120. Torres, N.V. S-system modelling approach to ecosystem: Application to a study of magnesium flow in a tropical forest. *Ecol. Model.* **1996**, *89*, 109–120. [CrossRef]

121. Sands, P.; Voit, E. Flux-based estimation of parameters in s-systems. *Ecol. Model.* **1996**, *93*, 75–88. [CrossRef]

122. Voit, E.O.; Sands, P.J. Modeling forest growth ii. Biomass partitioning in scots pine. *Ecol. Model.* **1996**, *86*, 73–89. [CrossRef]

123. Martin, P.-G. The use of canonical S-system modelling for condensation of complex dynamic models. *Ecol. Model.* **1997**, *103*, 43–70. [CrossRef]

124. Kaitaniemi, P. A canonical model of tree resource allocation after defoliation and bud consumption. *Ecol. Model.* **2000**, *129*, 259–272. [CrossRef]

125. Renton, M.; Kaitaniemi, P.; Hanan, J. Functional–structural plant modelling using a combination of architectural analysis, l-systems and a canonical model of function. *Ecol. Model.* **2005**, *184*, 277–298. [CrossRef]

126. Sorribas, A.; Hernandez-Bermejo, B.; Vilaprinyo, E.; Alves, R. Cooperativity and saturation in biochemical networks: A saturable formalism using Taylor series approximations. *Biotechnol. Bioeng.* **2007**, *97*, 1259–1277. [CrossRef] [PubMed]

127. Wu, L.; Wang, W.; van Winden, W.A.; van Gulik, W.M.; Heijnen, J.J. A new framework for the estimation of control parameters in metabolic pathways using lin-log kinetics. *Eur. J. Biochem.* **2004**, *271*, 3348–3359. [CrossRef] [PubMed]

128. Visser, D.; Heijnen, J.J. Dynamic simulation and metabolic re-design of a branched pathway using linlog kinetics. *Metab. Eng.* **2003**, *5*, 164–176. [CrossRef]

129. Heijnen, J.J. Approximative kinetic formats used in metabolic network modeling. *Biotechnol. Bioeng.* **2005**, *91*, 534–545. [CrossRef] [PubMed]

130. Del Rosario, R.C.H.; Mendoza, E.; Voit, E.O. Challenges in lin-log modelling of glycolysis in *Lactococcus lactis*. *IET Syst. Biol.* **2008**, *2*, 136–149. [CrossRef] [PubMed]

131. Wang, F.S.; Ko, C.L.; Voit, E.O. Kinetic modeling using S-systems and lin-log approaches. *Biochem. Eng. J.* **2007**, *33*, 238–247. [CrossRef]

132. Chou, I.C.; Voit, E.O. Estimation of dynamic flux profiles from metabolic time series data. *BMC Syst. Biol.* **2012**, *6*, 84. [CrossRef] [PubMed]

133. Dolatshahi, S.; Voit, E.O. Identification of metabolic pathway systems. *Front. Genet.* **2016**, *7*, 6. [CrossRef] [PubMed]

134. Iwata, M.; Shiraishi, F.; Voit, E.O. Coarse but efficient identification of metabolic pathway systems. *Int. J. Syst. Biol.* **2013**, *4*, 57.

135. Voit, E.O.; Goel, G.; Chou, I.C.; Fonseca, L.L. Estimation of metabolic pathway systems from different data sources. *Iet. Syst. Biol.* **2009**, *3*, 513–522. [CrossRef] [PubMed]

136. Hartmann, A.; Schreiber, F. Integrative analysis of metabolic models—From structure to dynamics. *Front. Bioeng. Biotechnol.* **2015**, *2*. [CrossRef] [PubMed]

137. Wu, Q.; Tian, T. Stochastic modeling of biochemical systems with multistep reactions using state-dependent time delay. *Sci. Rep.* **2016**, *6*, 31909. [CrossRef] [PubMed]

138. Yu, Y.; Dong, W.; Altimus, C.; Tang, X.; Griffith, J.; Morello, M.; Dudek, L.; Arnold, J.; Schüttler, H.-B. A genetic network for the clock of *Neurospora crassa*. *Proc. Natl. Acad. Sci. USA* **2007**, *104*, 2809–2814. [CrossRef] [PubMed]

139. Deng, Z.; Arsenault, S.; Caranica, C.; Griffith, J.; Zhu, T.; Al-Omari, A.; Schuttler, H.B.; Arnold, J.; Mao, L. Synchronizing stochastic circadian oscillators in single cells of *Neurospora crassa*. *Sci. Rep.* **2016**, *6*, 35828. [CrossRef] [PubMed]

140. Guerriero, M.L.; Akman, O.E.; van Ooijen, G. Stochastic models of cellular circadian rhythms in plants help to understand the impact of noise on robustness and clock structure. *Front. Plant Sci.* **2014**, *5*, 564. [CrossRef] [PubMed]

141. Guerriero, M.L.; Pokhilko, A.; Fernández, A.P.; Halliday, K.J.; Millar, A.J.; Hillston, J. Stochastic properties of the plant circadian clock. *J. R. Soc. Interface* **2012**, *9*, 744–756. [CrossRef] [PubMed]

142. Akman, O.E.; Ciocchetta, F.; Degasperi, A.; Guerriero, M.L. Modelling biological clocks with bio-pepa: Stochasticity and robustness for the *Neurospora crassa* circadian network. In Proceedings of the Computational Methods in Systems Biology: 7th International Conference (CMSB 2009), Bologna, Italy, 31 August–1 September 2009; Degano, P., Gorrieri, R., Eds.; Springer: Berlin, Heidelberg, 2009; pp. 52–67.

143. Gonze, D.; Halloy, J.; Goldbeter, A. Deterministic versus stochastic models for circadian rhythms. *J. Biol. Phys.* **2002**, *28*, 637–653. [CrossRef] [PubMed]

144. Sweetlove, L.J.; Fernie, A.R. The spatial organization of metabolism within the plant cell. *Annu. Rev. Plant Biol.* **2013**, *64*, 723–746. [CrossRef] [PubMed]

145. Grafahrend-Belau, E.; Junker, A.; Eschenroder, A.; Muller, J.; Schreiber, F.; Junker, B.H. Multiscale metabolic modeling: Dynamic flux balance analysis on a whole-plant scale. *Plant Physiol.* **2013**, *163*, 637–647. [CrossRef] [PubMed]

146. Davison, B.H.; Drescher, S.R.; Tuskan, G.A.; Davis, M.F.; Nghiem, N.P. Variation of s/g ratio and lignin content in a populus family influences the release of xylose by dilute acid hydrolysis. *Appl. Biochem. Biotechnol.* **2006**, *129–132*, 427–435. [CrossRef]

147. Van Acker, R.; Vanholme, R.; Storme, V.; Mortimer, J.C.; Dupree, P.; Boerjan, W. Lignin biosynthesis perturbations affect secondary cell wall composition and saccharification yield in *Arabidopsis thaliana*. *Biotechnol. Biofuels* **2013**, *6*, 46. [CrossRef] [PubMed]

148. Escamilla-Trevino, L.L.; Shen, H.; Uppalapati, S.R.; Ray, T.; Tang, Y.; Hernandez, T.; Yin, Y.; Xu, Y.; Dixon, R.A. Switchgrass (*Panicum virgatum*) possesses a divergent family of cinnamoyl coa reductases with distinct biochemical properties. *New Phytol.* **2010**, *185*, 143–155. [CrossRef] [PubMed]

149. Shen, H.; He, X.; Poovaiah, C.R.; Wuddineh, W.A.; Ma, J.; Mann, D.G.; Wang, H.; Jackson, L.; Tang, Y.; Stewart, C.N., Jr.; et al. Functional characterization of the switchgrass (*Panicum virgatum*) r2r3-myb transcription factor pvmyb4 for improvement of lignocellulosic feedstocks. *New Phytol.* **2012**, *193*, 121–136. [CrossRef] [PubMed]

150. Fu, C.; Mielenz, J.R.; Xiao, X.; Ge, Y.; Hamilton, C.Y.; Rodriguez, M., Jr.; Chen, F.; Foston, M.; Ragauskas, A.; Bouton, J.; et al. Genetic manipulation of lignin reduces recalcitrance and improves ethanol production from switchgrass. *Proc. Natl. Acad. Sci. USA* **2011**, *108*, 3803–3808. [CrossRef] [PubMed]

151. Torres, N.V.; Voit, E.O. *Pathway Analysis and Optimization in Metabolic Engineering*; Cambridge University Press: Cambridge, UK; New York, NY, USA, 2002; xiv, 305p.

152. Voit, E.O. *Computational Analysis of Biochemical Systems: A Practical Guide for Biochemists and Molecular Biologists*; Cambridge University Press: New York, NY, USA, 2000; xii, 531p.

processes

MDPI

Opinion

On the Use of Multivariate Methods for Analysis of Data from Biological Networks

Troy Vargason [1,2], Daniel P. Howsmon [2,3], Deborah L. McGuinness [4,5] and Juergen Hahn [1,2,3,*]

1 Department of Biomedical Engineering, Rensselaer Polytechnic Institute, Troy, NY 12180, USA; vargat@rpi.edu
2 Center for Biotechnology and Interdisciplinary Studies, Rensselaer Polytechnic Institute, Troy, NY 12180, USA; howsmd@rpi.edu
3 Department of Chemical and Biological Engineering, Rensselaer Polytechnic Institute, Troy, NY 12180, USA
4 Department of Computer Science, Rensselaer Polytechnic Institute, Troy, NY 12180, USA; dlm@cs.rpi.edu
5 Department of Cognitive Science, Rensselaer Polytechnic Institute, Troy, NY 12180, USA
* Correspondence: hahnj@rpi.edu; Tel.: +1-518-276-2138

Academic Editors: Rudiyanto Gunawan and Neda Bagheri
Received: 9 June 2017; Accepted: 30 June 2017; Published: 3 July 2017

Abstract: Data analysis used for biomedical research, particularly analysis involving metabolic or signaling pathways, is often based upon univariate statistical analysis. One common approach is to compute means and standard deviations individually for each variable or to determine where each variable falls between upper and lower bounds. Additionally, p-values are often computed to determine if there are differences between data taken from two groups. However, these approaches ignore that the collected data are often correlated in some form, which may be due to these measurements describing quantities that are connected by biological networks. Multivariate analysis approaches are more appropriate in these scenarios, as they can detect differences in datasets that the traditional univariate approaches may miss. This work presents three case studies that involve data from clinical studies of autism spectrum disorder that illustrate the need for and demonstrate the potential impact of multivariate analysis.

Keywords: multivariate statistics; Fisher discriminant analysis; probability density function; autism spectrum disorder; one carbon metabolism; transsulfuration; urine toxic metals; classification; machine learning

1. Introduction

Statistical analysis is a critical component for supporting any finding—whether from a clinical trial or other data collection. While there are numerous types of scenarios where such an analysis may need to be applied, two expository examples are: (1) when a clinical trial tests measurements from two or more populations, such as healthy versus diseased or placebo versus treatment; or (2) when a patient's blood sample is analyzed and the measured values are compared against reference ranges for a healthy individual. In both cases, the analysis is typically performed by comparing the representative value of one specific measured quantity against the same measured quantity of others, and this comparison is typically done for each measured quantity. However, such an approach will ignore correlations that may exist between the different measured quantities. If the measured quantities are representative of activity in a biological network where components are connected via reactions, interactions, or regulatory effects, such as in metabolic or signaling pathways, then traditional univariate approaches will potentially misrepresent the true behavior of the system under investigation. Multivariate analysis can address this shortcoming and, more accurately, it can be used to elucidate the characteristics of a biological network.

The value of considering multiple quantities simultaneously is recognized in the medical and biomedical communities, as demonstrated by the use of measurement ratios for univariate statistical analysis. The ratio of S-adenosylmethionine (SAM) to S-adenosylhomocysteine (SAH), for example, is used as an indicator of DNA methylation capacity [1]. Kidney functioning can be assessed with the ratio of blood urea nitrogen to creatinine in the plasma [2]. Furthermore, the ratio of total cholesterol to high-density lipoprotein cholesterol is used to provide an assessment of cardiovascular health [3]. Using ratios or observing the statistical distribution of ratios, instead of analyzing the separate values individually, can be advantageous as the interactions between different biological components may then be considered. However, the ability to take correlations of a larger number of measurements into account, without needing to specify the relationships, would be of even greater benefit. Multivariate statistical methods, such as Fisher Discriminant Analysis (FDA) [4] and its nonlinear extension, Kernel Fisher Discriminant Analysis (KFDA) [5], are promising options as they can address the aforementioned drawbacks of univariate analytical approaches.

This paper provides three case studies that compare the results obtained from univariate and multivariate statistical analyses of data from clinical studies. These case studies illustrate the benefits of using multivariate techniques over their univariate counterparts. While one must be careful when drawing conclusions from specific case studies about a more general setting, this work is nevertheless intended to highlight examples of advantages that can be gained by using multivariate analysis techniques, especially in cases where biological networks are involved.

2. Preliminary Information

2.1. Univariate Statistical Analysis

Univariate analyses are those aiming to summarize the characteristics of a single variable. These produce the statistics commonly reported in scientific literature, including the mean, standard deviation, and quantiles. When comparing a single measurement between two study populations, such as placebo and treatment groups, the two-sample *t*-test can be used to test for a significant difference in the group means, provided that the measurement is normally distributed in both groups [6]. Alternatively, the Mann-Whitney U test allows one to test for a significant difference in medians between two identical, but shifted, distributions [7].

2.2. Multivariate Statistical Analysis

Multivariate analysis involves the investigation of multiple variables simultaneously and encompasses a number of techniques that can be used to model data arising from complex systems. Such techniques take on a variety of forms and are used for a number of different tasks. For example, analyses of variance models [8] are commonly used to test the effects of multiple categorical factors on a measured response variable. The support vector machine [9] is a popular option for the supervised classification of groups of data consisting of a number of measurements. Additionally, hierarchical clustering [10] can be used for cluster analysis, while partial least squares regression [11] offers an approach for parameter estimation. Multivariate methods are often implemented in machine learning tasks in which models are developed with existing data and then used to predict new data. FDA is a useful method for maximizing separation between two or more groups of data samples [4] and is most appropriate when the input variables are continuous and normally distributed [12].

The input of FDA is a set of data samples X, where each sample x is a vector containing a fixed number of measurements. With a two-class problem (again consider the placebo versus treatment example), a subset of these samples X_1 belongs to one class while the remaining subset of samples X_2 belongs to the other class. The purpose of FDA is to calculate the projection vector w, which transforms each x to a single score variable t, that best separates the samples in X_1 and X_2. Separability is quantified by J, the ratio of the between-class scatter to the within-class scatter, and w is chosen to maximize this quantity [4]. Figure 1a summarizes this linear transformation performed in FDA as applied to individual samples.

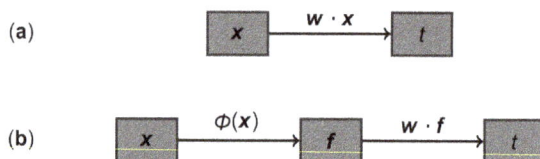

Figure 1. Schematics of the transformations used in Fisher Discriminant Analysis (FDA) and Kernel Fisher Discriminant Analysis (KFDA): (**a**) In FDA, the dot product of vector w with data sample x is calculated to obtain the projected value t; (**b**) KFDA first maps each sample x to a higher-dimensional space f according to the nonlinear transformation $\varphi(x)$. The dot product of w with f (rather than with x) is then calculated to obtain the projection t.

The principle of KFDA is similar to that of FDA, except that KFDA is capable of modeling nonlinear relationships between input variables rather than just linear ones. Before calculating a projection direction w to best separate X_1 and X_2, KFDA first applies a nonlinear transformation to each x, expressed as $f = \varphi(x)$, to map each to a higher-dimensional variable space f. Since the explicit mapping of $\varphi(x)$ is not known, an implicit mapping can be defined such that the inner product between any two $\varphi(x)$ is a Mercer kernel [5]. In a two-class problem, all f belonging to one class make up F_1 while the f in the other class comprise F_2. The vector w that best separates F_1 and F_2 is then determined, with the linear projection $t = w{\cdot}f$ capturing nonlinear relationships in the original variable space of x. Like FDA, nonlinear KFDA also aims to maximize the value of J. A schematic of the operations involved in KFDA is provided in Figure 1b. It should also be noted that the radial basis function, a commonly-used kernel, will be used in this work.

3. Advantages of Multivariate Approaches for Biological Network Analysis

Three case studies are presented in this section that illustrate some benefits of using multivariate approaches to analyze biological networks. The focus of these case studies is on folate-dependent one-carbon metabolism (FOCM) and transsulfuration (TS), two metabolic pathways with critical roles in the human body (Figure 2). FOCM, which occurs in every cell type [13], is involved with the epigenetic control of gene expression through DNA methylation [14]. The TS pathway, initiated by the conversion of homocysteine to cystathionine, is found in the liver, kidney, pancreas, small intestine, and brain, and contributes to the management of intracellular oxidative stress [15,16]. The FOCM and TS pathways are connected and together form an important juncture in the larger metabolic networks of human cells.

FOCM and TS are believed to be closely intertwined with genetic and environmental factors associated with autism spectrum disorder (ASD) predisposition [17] and therefore are often the focus of clinical studies investigating metabolic abnormalities in ASD [18–20]. These studies have found the ratio of S-adenosylmethionine (SAM) to S-adenosylhomocysteine (SAH) [21] to be reduced in individuals with ASD compared to neurotypical (NT) peers, which suggests a reduced DNA methylation capacity. The same studies have determined an increased proportion of oxidized to reduced glutathione [22], an important antioxidant, to indicate an irregular balance between oxidants and antioxidants (redox status) in ASD.

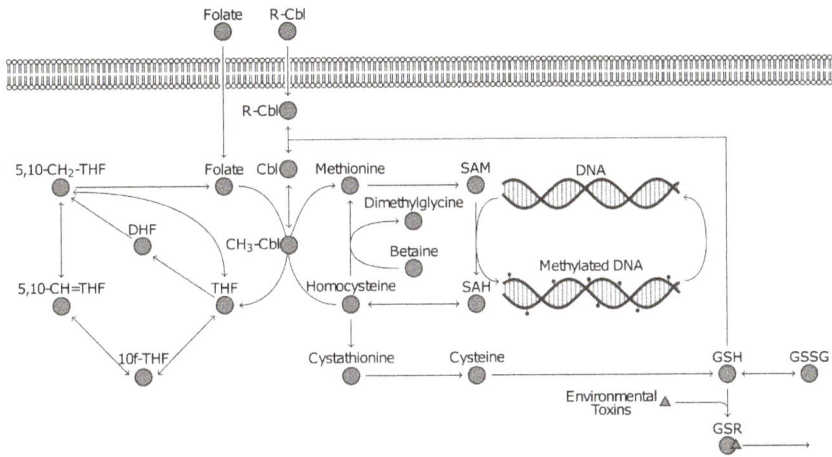

Figure 2. Diagram of major metabolites and reactions involved in the folate-dependent one-carbon metabolism (FOCM) and transsulfuration (TS) pathways. DNA methylation plays an important role in epigenetics and glutathione (GSH) is responsible for the clearance of environmental toxins.

The case studies that follow will highlight three unique aspects of multivariate analysis. First, the utility of incorporating multiple measurements for assessing network activity will be demonstrated using a general example. Second, advantages of using multivariate over univariate methods to analyze FOCM/TS metabolite data will be studied in the context of ASD classification. Third, the ability of nonlinear multivariate approaches to uncover relationships that linear analyses cannot describe will be explored, with a focus on measurements of toxic metals from the urine of individuals with ASD.

3.1. Advantages of Using Multiple Correlated Measurements for Diagnosis: A General Case

Consider a subset of reactions in FOCM associated with DNA methylation to be represented by the model in Figure 3. This model is taken to describe FOCM activity in liver cells. The metabolic reactions are assumed to proceed according to mass action kinetics and the reaction rates are thus proportional to the concentrations of the substrates, similar to the FOCM/TS model design used in a previous study [23]. In this model, methionine is delivered to liver cells at the rate v_{in}. Methionine is converted to SAM at a rate v_1 by methionine adenosyltransferase enzymes. SAM is then converted to SAH by methyltransferase enzymes at the rate v_2, or is depleted by other reactions and excreted at a rate described by $v_{deplete}$. Finally, SAH is converted to other FOCM products at a rate v_{out}.

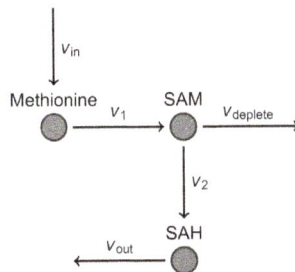

Figure 3. A simplified representation of a subset of reactions in FOCM responsible for DNA methylation.

Recall that a reduced SAM/SAH ratio has been observed in individuals with ASD and indicates a lowered capacity for DNA methylation. In the context of the metabolic model, this implies one of five scenarios: (1) reduced SAM and relatively normal SAH; (2) relatively normal SAM and elevated SAH; (3) both reduced SAM and elevated SAH; (4) elevated SAM and further elevated SAH; or (5) reduced SAH and further reduced SAM. In each scenario, the measurement of both SAM and SAH is required to make an informed assessment about DNA methylation capacity. Therefore, measuring SAM or SAH alone will not provide sufficient information to form meaningful conclusions about methylation status.

For example, suppose a patient has significantly increased v_{in}, which can be due to a number of reasons. All modeled metabolite concentrations (methionine, SAM, SAH) will then increase with time, along with their associated reaction rates. Clinical measurement of SAH sometime afterwards will indicate an elevated concentration of SAH, and following scenario (2) or scenario (3) the unwary clinician might conclude that the patient has a decreased SAM/SAH ratio. However, with an additional measurement of SAM it would be discovered that the SAM concentration is also elevated and the SAM/SAH ratio is relatively unchanged. The only way to verify this is to incorporate multiple measurements into the diagnosis and obtain a bigger picture of the network being studied.

A potential alternative to this multivariate approach would be to develop a comprehensive network model of the metabolic pathways under investigation and analyze the behavior of the network as a whole. While this can provide correlational (and sometimes causal) information that a multivariate statistical approach cannot, it also has several drawbacks. For one, a network model requires reasonably extensive knowledge of the network's structure and properties, which are not always known, or a very large dataset to construct the network's structure. Understanding the network's behavior then necessitates that the measurements be available for a large number of components of the network, whereas a multivariate analysis can be performed with just a subset of these measurements and without specifying the relationships between individual components. The presented multivariate approach thus offers a simplified, yet effective, representation of the network that can serve as a biomarker for the disorder or disease of interest.

3.2. Advantages of Using Multivariate Approaches over Univariate Approaches: Application to ASD Classification Using Clinical Measurements of FOCM/TS Metabolites

The purpose of this case study is to illustrate the benefit of incorporating multiple measurements, rather than a collection of individual ones, into a procedure for classifying two groups of data. To demonstrate this point, data from the Integrated Metabolic and Genomic Endeavor (IMAGE) study at Arkansas Children's Hospital Research Institute [20] will be used. The IMAGE study investigates plasma profiles of FOCM and TS metabolites in individuals with ASD and how they compare to those of NT individuals. Measurements of primary interest in this study are methionine cycle and TS metabolites, as well as DNA methylation and oxidative stress markers.

ASD classification has been performed with high accuracy by applying FDA to measurements from the IMAGE study [24]. In a multivariate analysis of these data, a subset of five measurements was found to provide excellent classification of the ASD and NT cohorts. These measurements, which are explained elsewhere in greater detail [20], were: (1) the percentage of DNA that is methylated (% DNA methylation), an indicator of epigenetic activity; (2) the concentration of 8-hydroxyguanosine, a marker of oxidative damage in DNA; (3) the concentration of glutamylcysteine, the precursor for glutathione; (4) the ratio of free oxidized cysteine to free reduced cysteine (free cystine/free cysteine), an indicator of extracellular redox status; and (5) the percentage of glutathione molecules that are oxidized (% oxidized glutathione).

Table 1 provides descriptive statistics for each of these measurements in the ASD and NT cohorts, along with p-values from the two-tailed Welch's t-test (significance level $\alpha = 0.05$). These numbers indicate a significant difference in the mean between the cohorts for all five measurements. To further characterize these differences, the probability density functions (PDFs) of each variable were plotted for each group (Figure 4). The differences in means between cohorts are apparent in these distributions.

However, there is still significant overlap of the PDFs, suggesting that these measurements will not allow for an accurate classification of a patient when considered individually.

Table 1. Means and standard deviations of five FOCM/TS measurements for the autism spectrum disorder (ASD) and neurotypical (NT) cohorts from the Integrated Metabolic and Genomic Endeavor (IMAGE) study. Reported *p*-values were obtained from the two-tailed Welch's *t*-test.

Measurement	ASD Mean ± SD $n = 83$	NT Mean ± SD $n = 76$	*p*-Value
% DNA methylation	3.37 ± 0.87	4.26 ± 0.90	<0.001
8-hydroxyguanosine (pmol/mg DNA)	89.2 ± 27.9	56.7 ± 17.9	<0.001
glutamylcysteine (μM)	1.87 ± 0.46	2.37 ± 0.59	<0.001
free cystine/free cysteine	1.51 ± 0.58	1.06 ± 0.35	<0.001
% oxidized glutathione	0.22 ± 0.07	0.12 ± 0.04	<0.001

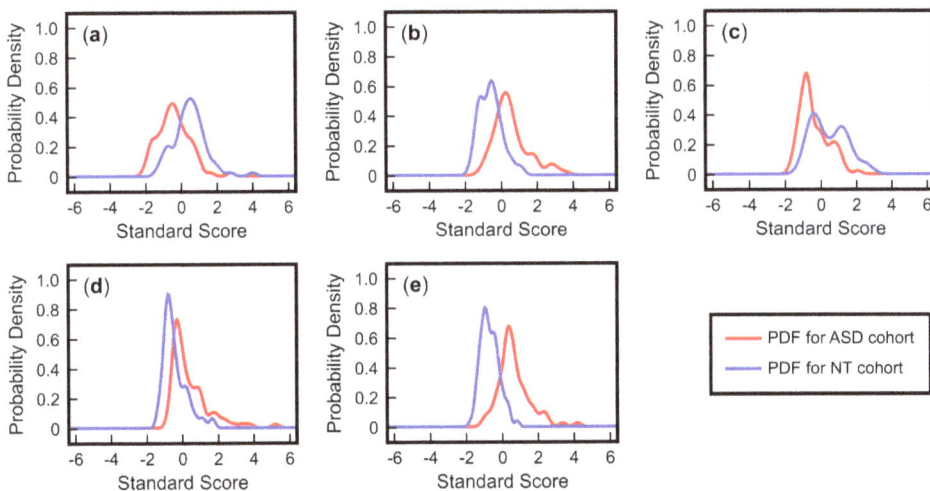

Figure 4. Probability density functions (PDFs) of five measurements for ASD and NT cohorts from the IMAGE study: (**a**) % DNA methylation; (**b**) 8-hydroxyguanosine; (**c**) glutamylcysteine; (**d**) free cystine/free cysteine; (**e**) % oxidized glutathione. These PDFs are based on the standardized values of each measurement (i.e., all samples for a measurement are scaled such that the mean value is 0 and the standard deviation is 1).

The use of multivariate methods such as FDA can address this issue. Figure 5 shows the results of applying FDA to these five measurements using leave-one-out cross-validation [25]; this method provides an independent assessment of the model performance by training the FDA model on all samples but one, obtaining a projected score for the left-out sample, and then repeating this process such that every sample has been left out exactly once. The resulting PDFs for the ASD and NT cohorts are well-separated, and when the indicated threshold is used for classification, the corresponding Type I and Type II errors are only 4.8% and 5%, respectively. It must be emphasized that since cross-validation was used in this analysis, the problem of potentially overfitting the FDA model by including more variables was addressed; these results also indicate the model's ability to accurately predict new data points that were not originally used to develop the model.

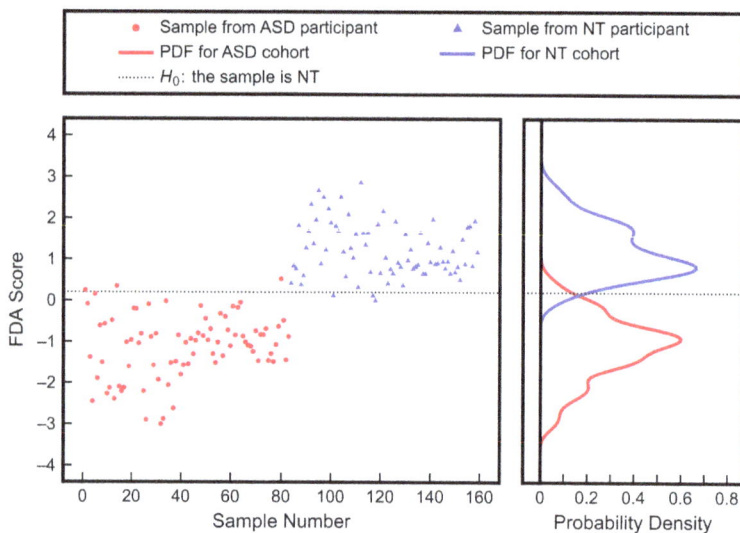

Figure 5. Multivariate analysis with FDA using five measurements from the IMAGE study (% DNA methylation, 8-hydroxyguanosine, glutamylcysteine, free cystine/free cysteine, and % oxidized glutathione). The scores are the projected values obtained by leave-one-out cross-validation with FDA, while the PDFs were obtained by fitting to the scores. The shown threshold corresponds to a Type I error of 4.8% and a Type II error of 5%.

In summary, univariate analysis of the five FOCM/TS measurements indicates significant differences in the means between the ASD and NT cohorts for each of the measurements. However, due to the variance in the measurements, these differences are not sufficiently large for purposes of classification. On the other hand, the application of a multivariate technique (in this case, FDA) allows us to simultaneously consider all of these measurements and determine a pattern in the data that can accurately predict if measurements come from a participant in the ASD or NT cohort.

3.3. Advantages of Nonlinear Approaches over Linear Approaches: Application to ASD Classification Using Clinical Measurements of Urine Toxic Metals

This final case study examines how nonlinear multivariate methods can uncover relationships among measurements that linear methods are unable to capture. The advantages of these nonlinear approaches have previously been shown using measurements of urine toxic metals that were collected as part of the Comprehensive Nutritional and Dietary Intervention Study at Arizona State University [26]. These data are again considered here.

Recall that the TS pathway is responsible for the synthesis of glutathione, which plays a major role in the regulation of oxidative stress. One use of glutathione is to aid with the removal of unwanted substances, such as toxic metals, from the body by binding them and subsequently facilitating excretion. Most of the excretion is done via feces [27], although other routes such as excretion via urine can also play a role [28]. Given that children with ASD have been found to have reduced levels of glutathione [18], it is likely that their toxic metal excretions will be different from those of their neurotypical peers. Thus, urine toxic metals can potentially be used as an indicator of FOCM and TS abnormalities in patients with ASD.

Descriptive univariate statistics for measurements of three urine toxic metals collected in the Comprehensive Nutritional and Dietary Intervention Study [26] are given in Table 2. It should be noted that each measurement is normalized by the amount of creatinine to address the varying dilution of each urine sample. Among these urine toxic metals, none had means that were significantly different between the ASD and NT cohorts when evaluated with the two-tailed Welch's *t*-test (significance level $\alpha = 0.05$). This univariate analysis suggests little to no separability between the ASD and NT groups based on these three measurements.

Table 2. Means and standard deviations of levels of three urine toxic metals in the ASD and NT cohorts from the Comprehensive Nutritional and Dietary Intervention Study. Metal levels are in units of μg/g of creatinine. Reported *p*-values were obtained from the two-tailed Welch's *t*-test.

Measurement	ASD Mean ± SD $n = 67$	NT Mean ± SD $n = 50$	*p*-Value
Aluminum	9.03 ± 6.55	8.55 ± 11.15	n.s.
Cesium	4.03 ± 1.92	3.74 ± 1.75	n.s.
Tungsten	0.29 ± 0.25	0.29 ± 0.21	n.s.

Applying FDA to these data does not produce any substantial separation between cohorts either (Figure 6). The PDFs resulting from leave-one-out cross-validation overlap almost entirely, with the corresponding Type I error at 50% and Type II error also at 50%. Using a linear multivariate approach thus does not offer any additional insights for classification. This is not unexpected, as the results of the univariate analysis also showed minimal differences between the ASD and NT measurements. However, there may be nonlinear relationships present that neither univariate nor linear multivariate techniques can describe.

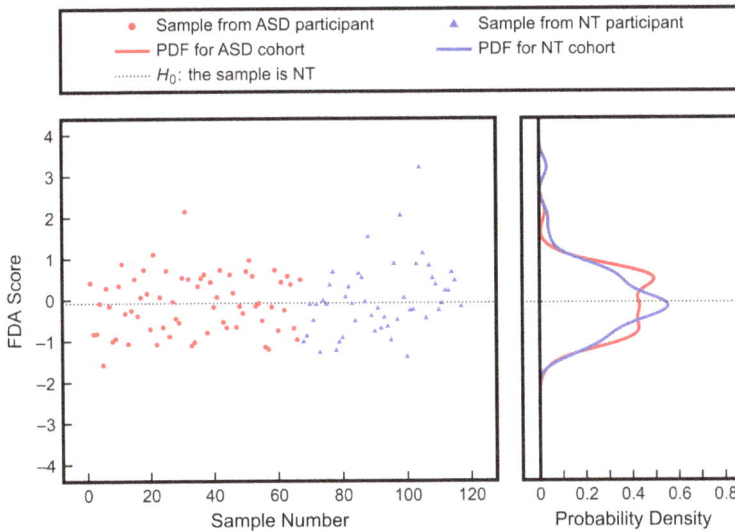

Figure 6. Results of classification using linear FDA with three urine toxic metal measurements (aluminum, cesium, tungsten) as inputs. FDA scores were from leave-one-out cross-validation and the PDFs were obtained by fitting to the scores. The Type I and Type II errors are both 50%.

Using nonlinear KFDA with these three urine toxic metal measurements improves the classification significantly, as seen in Figure 7. The PDFs after leave-one-out cross-validation with KFDA produce Type I and Type II errors of 29% and 28%, respectively. These results are notably better than those obtained from the linear analysis, though still far from being usable as a diagnostic tool. This inability to accurately classify the two cohorts highlights that KFDA will not detect strong differences between groups of data that are very similar, as is the case with the three urine toxic metal measurements presented here. It is nevertheless important to note that the nonlinear approach was still able to identify certain differences in the patterns in the data between groups that the linear analysis missed. This example highlights that a univariate or a linear approach being unable to find differences between two groups does not mean that differences may not exist. This is especially so for more complex relationships between variables that may be present in biological networks.

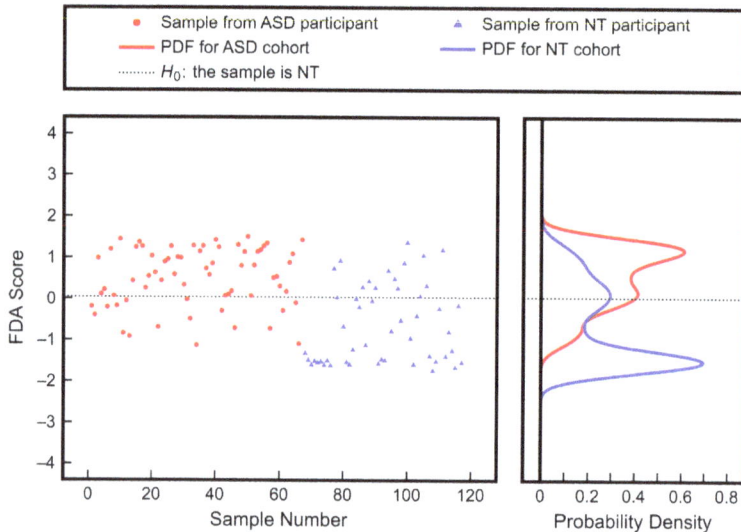

Figure 7. Results of classification using nonlinear KFDA with three urine toxic metal measurements (aluminum, cesium, tungsten) as inputs. KFDA scores were from leave-one-out cross-validation and the PDFs were obtained by fitting to the scores. The corresponding Type I and Type II errors are 29% and 28%, respectively.

4. Conclusions

Statistical analysis is an integral part of any clinical trial and is also critical for evaluating medical laboratory test results. While the current state of practice in many areas of biomedical research involving metabolic or signaling pathways is to use univariate statistical analysis to evaluate one measurement at a time (across a cohort where this is applicable), this approach is sub-optimal when the measured quantities are correlated in some form, as is the case when they are connected via a biological network. This work included three case studies involving clinical data to demonstrate that significant advantages can be gained from using multivariate statistical analysis on these types of data. It is the opinion of the authors that multivariate analysis techniques should be more broadly considered for measurements taken from biological networks.

Acknowledgments: The authors gratefully acknowledge partial financial support from the National Institutes of Health (https://www.nih.gov/, Grant 1R01AI110642). Additionally, the authors would like to thank S. Jill James (University of Arkansas for Medical Sciences) for providing the FOCM/TS data and James B. Adams (Arizona State University) for providing the urine toxic metal data used in this work.

Author Contributions: J.H. and D.L.M. conceived of the study; T.V., D.P.H. and J.H. formulated the case studies; T.V. and D.P.H. analyzed the data and created the visuals; all authors aided in the writing and revision of the paper.

Conflicts of Interest: The authors declare no conflict of interest.

References

1. Frye, R.E.; James, S.J. Metabolic pathology of autism in relation to redox metabolism. *Biomark. Med.* **2014**, *8*, 321–330. [CrossRef] [PubMed]
2. Morgan, D.B.; Carver, M.E.; Payne, R.B. Plasma creatinine and urea: Creatinine ratio in patients with raised plasma urea. *Br. Med. J.* **1977**, *2*, 929–932. [CrossRef] [PubMed]
3. Lemieux, I.; Lamarche, B.; Couillard, C.; Pascot, A.; Cantin, B.; Bergeron, J.; Dagenais, G.R.; Després, J.-P. Total cholesterol/HDL cholesterol ratio vs LDL cholesterol/HDL cholesterol ratio as indices of ischemic heart disease risk in men: The Quebec Cardiovascular Study. *Arch. Intern. Med.* **2001**, *161*, 2685–2692. [CrossRef] [PubMed]
4. Fisher, R.A. The use of multiple measurements in taxonomic problems. *Ann. Eugen.* **1936**, *7*, 179–188. [CrossRef]
5. Mika, S.; Ratsch, G.; Weston, J.; Scholkopf, B.; Mullers, K.R. Fisher discriminant analysis with kernels. In Proceedings of the 1999 IEEE Neural Networks for Signal Processing IX Workshop, Madison, WI, USA, 23–25 August 1999; pp. 41–48.
6. Ruxton, G.D. The unequal variance *t*-test is an underused alternative to Student's *t*-test and the Mann–Whitney *U* test. *Behav. Ecol.* **2006**, *17*, 688–690. [CrossRef]
7. Mann, H.B.; Whitney, D.R. On a test of whether one of two random variables is stochastically larger than the other. *Ann. Math. Stat.* **1947**, *18*, 50–60. [CrossRef]
8. Scheffé, H. *The Analysis of Variance*; John Wiley & Sons: New York, NY, USA, 1999.
9. Cortes, C.; Vapnik, V. Support-vector networks. *Mach. Learn.* **1995**, *20*, 273–297. [CrossRef]
10. Johnson, S.C. Hierarchical clustering schemes. *Psychometrika* **1967**, *32*, 241–254. [CrossRef] [PubMed]
11. Wold, S.; Sjöström, M.; Eriksson, L. PLS-regression: A basic tool of chemometrics. *Chemom. Intell. Lab. Syst.* **2001**, *58*, 109–130. [CrossRef]
12. Hastie, T.; Tibshirani, R.; Friedman, J. *The Elements of Statistical Learning: Data Mining, Inference, and Prediction, Second Edition*, 2nd ed.; Springer: New York, NY, USA, 2011.
13. Appling, D.R. Compartmentation of folate-mediated one-carbon metabolism in eukaryotes. *FASEB J.* **1991**, *5*, 2645–2651. [PubMed]
14. Anderson, O.S.; Sant, K.E.; Dolinoy, D.C. Nutrition and epigenetics: An interplay of dietary methyl donors, one-carbon metabolism and DNA methylation. *J. Nutr. Biochem.* **2012**, *23*, 853–859. [CrossRef] [PubMed]
15. Finkelstein, J.D.; Martin, J.J. Homocysteine. *Int. J. Biochem. Cell Biol.* **2000**, *32*, 385–389. [CrossRef]
16. Vitvitsky, V.; Thomas, M.; Ghorpade, A.; Gendelman, H.E.; Banerjee, R. A functional transsulfuration pathway in the brain links to glutathione homeostasis. *J. Biol. Chem.* **2006**, *281*, 35785–35793. [CrossRef] [PubMed]
17. Deth, R.; Muratore, C.; Benzecry, J.; Power-Charnitsky, V.-A.; Waly, M. How environmental and genetic factors combine to cause autism: A redox/methylation hypothesis. *NeuroToxicology* **2008**, *29*, 190–201. [CrossRef] [PubMed]
18. James, S.J.; Melnyk, S.; Jernigan, S.; Cleves, M.A.; Halsted, C.H.; Wong, D.H.; Cutler, P.; Bock, K.; Boris, M.; Bradstreet, J.J.; et al. Metabolic endophenotype and related genotypes are associated with oxidative stress in children with autism. *Am. J. Med. Genet. B Neuropsychiatr. Genet.* **2006**, *141*, 947–956. [CrossRef] [PubMed]
19. Adams, J.B.; Audhya, T.; McDonough-Means, S.; Rubin, R.A.; Quig, D.; Geis, E.; Gehn, E.; Loresto, M.; Mitchell, J.; Atwood, S.; et al. Nutritional and metabolic status of children with autism vs. neurotypical children, and the association with autism severity. *Nutr. Metab.* **2011**, *8*, 34. [CrossRef] [PubMed]
20. Melnyk, S.; Fuchs, G.J.; Schulz, E.; Lopez, M.; Kahler, S.G.; Fussell, J.J.; Bellando, J.; Pavliv, O.; Rose, S.; Seidel, L.; et al. Metabolic imbalance associated with methylation dysregulation and oxidative damage in children with autism. *J. Autism Dev. Disord.* **2012**, *42*, 367–377. [CrossRef] [PubMed]
21. Yi, P.; Melnyk, S.; Pogribna, M.; Pogribny, I.P.; Hine, R.J.; James, S.J. Increase in plasma homocysteine associated with parallel increases in plasma S-adenosylhomocysteine and lymphocyte DNA hypomethylation. *J. Biol. Chem.* **2000**, *275*, 29318–29323. [CrossRef] [PubMed]
22. Jones, D.P. Redox potential of GSH/GSSG couple: Assay and biological significance. *Methods Enzymol.* **2002**, *348*, 93–112. [PubMed]

23. Vargason, T.; Howsmon, D.P.; Melnyk, S.; James, S.J.; Hahn, J. Mathematical modeling of the methionine cycle and transsulfuration pathway in individuals with autism spectrum disorder. *J. Theor. Biol.* **2017**, *416*, 28–37. [CrossRef] [PubMed]

24. Howsmon, D.P.; Kruger, U.; Melnyk, S.; James, S.J.; Hahn, J. Classification and adaptive behavior prediction of children with autism spectrum disorder based upon multivariate data analysis of markers of oxidative stress and DNA methylation. *PLoS Comput. Biol.* **2017**, *13*, e1005385. [CrossRef] [PubMed]

25. Kohavi, R. A study of cross-validation and bootstrap for accuracy estimation and model selection. In Proceedings of the 14th International Joint Conference on Artificial Intelligence, Montreal, QC, Canada, 20–25 August 1995; Morgan Kaufmann Publishers Inc.: San Francisco, CA, USA, 1995; Volume 2, pp. 1137–1143.

26. Adams, J.B.; Howsmon, D.P.; Kruger, U.; Geis, E.; Gehn, E.; Fimbres, V.; Pollard, E.; Mitchell, J.; Ingram, J.; Hellmers, R.; et al. Significant association of urinary toxic metals and autism-related symptoms—A nonlinear statistical analysis with cross validation. *PLoS ONE* **2017**, *12*, e0169526. [CrossRef] [PubMed]

27. Adams, J.B.; Audhya, T.; McDonough-Means, S.; Rubin, R.A.; Quig, D.; Geis, E.; Gehn, E.; Loresto, M.; Mitchell, J.; Atwood, S.; et al. Toxicological status of children with autism vs. neurotypical children and the association with autism severity. *Biol. Trace Elem. Res.* **2012**, *151*, 171–180. [CrossRef] [PubMed]

28. Rossignol, D.A.; Genuis, S.J.; Frye, R.E. Environmental toxicants and autism spectrum disorders: A systematic review. *Transl. Psychiatry* **2014**, *4*, e360. [CrossRef] [PubMed]

processes

MDPI

Article

Characterizing Gene and Protein Crosstalks in Subjects at Risk of Developing Alzheimer's Disease: A New Computational Approach

Kanchana Padmanabhan [1,2], Kelly Nudelman [3], Steve Harenberg [1], Gonzalo Bello [1], Dongwha Sohn [1,2], Katie Shpanskaya [4], Priyanka Tiwari Dikshit [1], Pallavi S. Yerramsetty [1], Rudolph E. Tanzi [5], Andrew J. Saykin [3], Jeffrey R. Petrella [6], P. Murali Doraiswamy [7] and Nagiza F. Samatova [1,2,*] for the Alzheimer's Disease Neuroimaging Initiative [†]

[1] Department of Computer Science, North Carolina State University, Raleigh, NC 27695, USA;
    kanchanapadmanabhan@gmail.com (K.P.); sdharenb@ncsu.edu (S.H.); gabellol@ncsu.edu (G.B.);
    dsohn@ncsu.edu (D.S.); ptdikshi@ncsu.edu (P.T.D.); pallaviyerramsetty98@gmail.com (P.S.Y.)
[2] Computer Science and Mathematics Division, Oak Ridge National Laboratory, Oak Ridge, TN 37831, USA
[3] Indiana Alzheimer Disease Center and the Center for Neuroimaging, Department of Radiology and Imaging
    Sciences, Indiana University School of Medicine, Indianapolis, IN 46202, USA; kholohan@iupui.edu (K.N.);
    asaykin@iupui.edu (A.J.S.)
[4] Department of Radiology, Stanford University School of Medicine, Stanford, CA 94025, USA;
    kss@stanford.edu
[5] Genetics and Aging Research Unit and Department of Neurology, Massachusetts General Hospital and
    Harvard Medical School Stanford University School of Medicine, Stanford, CA 02129, USA;
    tanzi@helix.mgh.harvard.edu
[6] Department of Radiology, Duke University Medical Center, Durham, NC 27710, USA;
    jeffrey.petrella@duke.edu
[7] Neurocognitive Disorders Program, Department of Psychiatry and the Duke Institute for Brain Sciences,
    Duke University Health System, Durham, NC 27710, USA; murali.doraiswamy@duke.edu
[*] Correspondence: samatova@csc.ncsu.edu; Tel.: +1-919-513-7575
[†] Data used in preparation of this article were obtained from the Alzheimer's disease Neuroimaging Initiative
    (ADNI) database (adni.loni.usc.edu). As such, the investigators within the ADNI contributed to the design
    and implementation of ADNI and/or provided data but did not participate in analysis or writing of this
    report. A complete listing of ADNI investigators can be found at:
    http://adni.loni.usc.edu/wp-content/uploads/how_to_apply/ADNI_Acknowledgement_List.pdf.

Received: 29 June 2017; Accepted: 13 August 2017; Published: 17 August 2017

Abstract: Alzheimer's disease (AD) is a major public health threat; however, despite decades of research, the disease mechanisms are not completely understood, and there is a significant dearth of predictive biomarkers. The availability of systems biology approaches has opened new avenues for understanding disease mechanisms at a pathway level. However, to the best of our knowledge, no prior study has characterized the nature of pathway crosstalks in AD, or examined their utility as biomarkers for diagnosis or prognosis. In this paper, we build the first computational crosstalk model of AD incorporating genetics, antecedent knowledge, and biomarkers from a national study to create a generic pathway crosstalk reference map and to characterize the nature of genetic and protein pathway crosstalks in mild cognitive impairment (MCI) subjects. We perform initial studies of the utility of incorporating these crosstalks as biomarkers for assessing the risk of MCI progression to AD dementia. Our analysis identified Single Nucleotide Polymorphism-enriched pathways representing six of the seven Kyoto Encyclopedia of Genes and Genomes pathway categories. Integrating pathway crosstalks as a predictor improved the accuracy by 11.7% compared to standard clinical parameters and apolipoprotein E ε4 status alone. Our findings highlight the importance of moving beyond discrete biomarkers to studying interactions among complex biological pathways.

Processes **2017**, *5*, 47

Keywords: pathway crosstalk; Alzheimer's disease; biomarker; disease prediction

1. Introduction

It is common knowledge that the prognostics of diseases such as Alzheimer's disease (AD) is of national importance. AD alone affects about 10% of the population over 65 years old [1,2], and is among the leading causes of death in patients over 75 years of age in the U.S. [3]. There is evidence suggesting that the progression to AD dementia begins years before it is clinically determined and is preceded by a phase of mild cognitive impairment (MCI), during which AD-related treatments are likely to be more effective. Thus, it is important to discover the mechanisms underlying risk of AD and to develop accurate biomarkers that reflect the complexity of the disease at an individual level. Although a number of biomarkers are currently being evaluated for use to predict AD or study disease progression (e.g., tau, *p*-tau181P, *β*-amyloid1-42, apolipoprotein E *ε*4 (*APOE ε*4), and microRNAs) [4–7], none of these markers are yet fully validated or approved for predicting the risk of AD. Indeed, AD is no longer seen as a disease of single discrete lesions, but as a perturbation of altered cortical networks by pathological processes in interlinked pathways. Hence, the application of systems biology methods to the discovery and characterization of novel biomarkers [8–20] has taken on greater promise and urgency.

The cellular mechanisms underlying many neurological disorders are complex, with crosstalks between multiple molecular pathways likely contributing to disease initiation and progression. In living organisms, pathways are said to crosstalk if they are linked together to perform biological functions as a system. Crosstalks can also be defined as interactions between signal transduction pathways, and usually take the form of protein or transmembrane interactions. A number of potential crosstalks have been noted in vitro in AD, such as those between amyloid and tau pathways, oxidative phosphorylation, the p53 signaling pathway, and apoptosis [21–23]. Another example is the reported crosstalk among MAPK, insulin, and calcium signaling pathways [24]. There is also evidence of crosstalk among pathways involved in the regulation of glycolysis metabolism, pathways involved in the regulation of the actin cytoskeleton, and apoptosis [24]. The latter crosstalk is also associated with other neurodegenerative disorders, such as Huntington disease and amyotrophic lateral sclerosis [24]. Furthermore, the cellular signaling pathways in AD have been reported, such as *Wnt* signaling, 5′ adenosine monophosphate-activated protein kinase, mammalian target of rapamycin, Sirtuin 1, and peroxisome proliferator-activated receptor gamma co-activator 1-α, and possible crosstalk between these pathways has been discussed [25]. For a review of multiple interacting pathways in neurodegenerative disease, see [26]. In clinical AD research studies of diagnosis or prognosis, biomarkers are typically treated as discrete entities, in part because biological pathway crosstalks between genes or proteins have not yet been fully characterized at a systems biology level in AD.

From the computational methodology standpoint, the study of pathway crosstalks is still in its infancy. Existing methods predict crosstalks between known metabolic pathways using chemical protein interaction networks [24,27–29]. However, these computational methods do not take advantage of the different chemical evidence available, such as direct binding, the biochemical evidence, such as phosphorylation, and the functional evidence, such as transcriptional regulation. Moreover, the discovery, characterization, and utilization of pathway crosstalks as biomarkers for disease prognosis has not been investigated.

Here, we use clinical, cognitive, and genetic data from a national cohort study, the Alzheimer's Disease Neuroimaging Initiative (ADNI-1), along with a systematic computational methodology to discover and characterize biological pathway crosstalks in subjects with MCI. We further examine the utility of these novel biomarkers to discriminate stable MCI from those who progress to AD dementia. The first part of the methodology (Figure 1), focuses on utilizing several existing evidence, such as chemical interaction, genetic interaction, domain interaction, and transcription factors, to identify potential pathway crosstalks. In the second part (Figure 2), Single Nucleotide Polymorphisms (SNPs) are used to find patient-specific pathway crosstalks as biomarkers. In the third part, we build and test initial prognostic models that use pathway crosstalks as biomarkers to predict patient progression from MCI progression to AD dementia (see Results). To the best of our knowledge, this is the first such systematic characterization of biological pathway crosstalk biomarkers associated with the risk of AD.

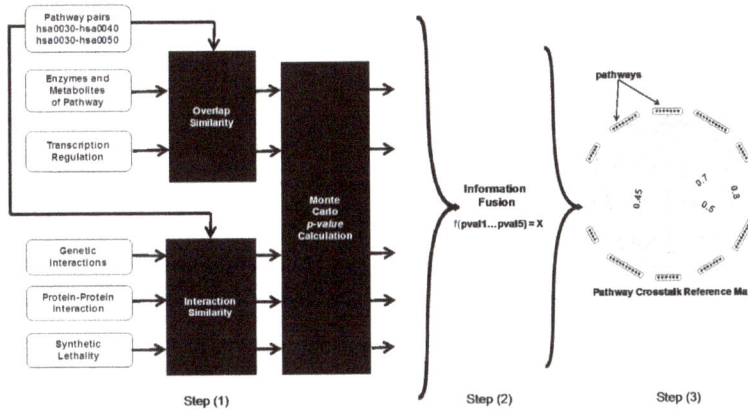

Figure 1. Identification of potential pathway crosstalks. The methodology has three steps: (1) quantifying crosstalk likelihood using multiple individual evidence to score each pathway pair, (2) obtaining a combined score using information fusion, and (3) building the crosstalk reference map.

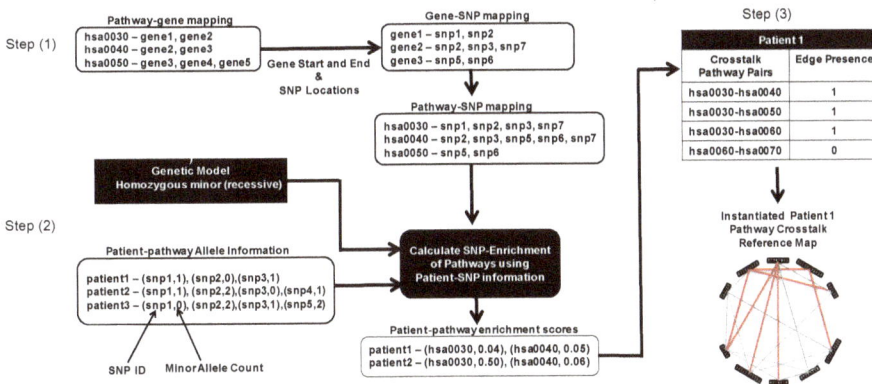

Figure 2. Identification of patient-specific pathway crosstalks. The methodology has three steps: (1) mapping the Single Nucleotide Polymorphisms (SNPs) to genes and in turn to pathways using the SNP and gene location information, (2) choosing a genetic model and calculating a patient-specific SNP enrichment score for each pathway using the patient's allele information, and (3) overlaying the pathway enrichment scores on the reference crosstalk map to build patient-specific pathway crosstalk maps.

2. Materials and Methods

Our methodology consists of the following steps: (A) identifying potential pathway crosstalks by using existing gene and protein data (Figure 1), (B) identifying patient-specific pathway crosstalks via SNP information (Figure 2), and (C) identifying significant pathway crosstalks as biomarkers for MCI progression to AD dementia progression prediction.

2.1. Identification of Potential Pathway Crosstalks

We quantify how likely it is that a pair of pathways will crosstalk based on biological datasets that provide evidence for possible crosstalks (including chemical interaction, genetic interaction, and transcription factors). To have a more robust pathway crosstalk map, we incorporate a wide array of evidence. The scores from each of these evidence are then combined to build one generic pathway crosstalk reference map analogous to the "Kyoto Encyclopedia of Genes and Genomes" (KEGG) pathway reference map.

The likelihood of a pathway pair crosstalking can be scored by utilizing one of two different methods. The first method is based on the presence of common elements, such as kinases and enzymes. The second method is based on the presence of interacting elements, such as chemically interacting proteins. In the following sections, we will discuss the different evidence used and their corresponding scoring methods.

2.1.1. Scoring Pathway Crosstalks Based on Common Elements

The pathway pairs were scored for how likely they are to crosstalk based on common elements from each of the following evidence:

- Shared enzymes and metabolites: The number of enzymes and metabolites shared by a pair of pathways is utilized as one of the evidence to identify potential pathway crosstalks. This is reasonable because a variation in the concentration of common enzymes or metabolites will affect both pathways.
- Phosphorylation: Phosphorylation, performed by protein kinases, is the addition of a phosphate group to a protein, which results in a change of the protein's function. Co-phosphorylated proteins in different pathways suggest potential pathway crosstalks.
- Transcriptional regulation: Genes with common transcription factors are likely coexpressed. Coexpressed genes in different pathways provide an avenue for the pathways to crosstalk. For each pathway pair, we find the group of transcription factors that have coexpressed genes in both pathways.

For each pair of pathways, P_i and P_j, we define the scoring function as Equation (1):

$$Overlap_{score}\left(P_i, P_j\right) = \frac{|Y(P_i) \cap Y(P_j)|}{min\left(|Y(P_i)|, |Y(P_j)|\right)}, \tag{1}$$

where $Y(P_i)$ is the set of proteins (enzymes, metabolites, transcription factors, kinases) associated with pathway P_i.

2.1.2. Scoring Pathway Crosstalks Based on Interacting Elements

The pathway pairs were scored for how likely they are to crosstalk based on interacting elements from each of the following evidence:

- Chemical interactions: Protein interactions have previously been used to identify pathway crosstalks [24,30]. Chemical interaction between proteins belonging to different pathways provides a mechanism for pathways to crosstalk.
- Genetic interactions: The use of genetic interactions for identifying pathway crosstalks stems from the concept of "between-pathway" interactions. This essentially states that if there is a genetic interaction between pathways, one pathway covers for the defects in the other pathway.
- Protein domain: Protein function is closely related to fundamental units of protein structure called "domains". In the domain interaction network, a pair of proteins has an edge if they are associated with the same set of protein domains. These edges are taken into consideration to assess for potential pathway crosstalks because of the common domains.

- Synthetically lethal gene pairs: Gene pairs whose simultaneous low- or non-expression can cause the organism to die are called synthetically lethal pairs [31,32]. The presence of synthetically lethal pairs of genes across two pathways is a possible sign of pathway crosstalks.

For each pair of pathways, P_i and P_j, we define the scoring function as Equation (2):

$$Interaction_{score}(P_i, P_j) = \frac{N_{inter}(P_i, P_j)}{|Y(P_i)| * |Y(P_j)|}, \tag{2}$$

where $N_{inter}(P_i, P_j)$ is the number of interactions (genetic, chemical, domain, synthetically lethal) that exist among the proteins associated with pathway P_i and the proteins associated with pathway P_j.

2.1.3. Significance Estimation of Pathway Crosstalk Scores

Estimating p-values using Monte Carlo methods [33] is a robust technique for statistical significance assessments. This technique was utilized to assess the significance of the scores obtained for the pathway crosstalks using different evidence, as follows:

1. For each pair of pathways, a score for how likely they are to crosstalk is calculated based on each evidence.
2. Each pathway is randomized by replacing all proteins in that pathway with randomly selected proteins from the set of all proteins in the organism. This pathway randomization step is repeated $W = 1000$ times, i.e., we obtain W sets of pathways with randomized proteins.
3. The evidence-specific scores for each pathway pair are recalculated W times using each set of pathways with randomized proteins.
4. An evidence-specific p-value is estimated for each pathway pair as R/W, where R is the number of randomized versions of that pathway pair that produce an evidence-specific score greater than or equal to the score obtained for the original pathway pair.

2.1.4. Combining the Scores for Each Pathway Crosstalk

For each pathway pair, we combine the evidence-specific p-values obtained using Monte Carlo methods. This gives a combined estimation for crosstalk likelihood between the pathway pair. To combine the p-values, we use the QFAST information fusion methodology proposed by Bailey and Gribskov [34], which is based on a theorem by Feller [35]. The QFAST methodology uses the product of the individual p-values as a test statistic to calculate the combined p-value; using the product of p-values as a test statistic has been shown to be a desirable method for information fusion [34]. One issue to consider is that some pathway pairs may not be scored by some of the evidence due to missing data. For those cases, we assign a p-value of 1 to denote that the particular evidence offers no information about those pathways crosstalking. The QFAST formula to calculate the combined p-value is Equation (3):

$$\left(\prod_{i=1}^{n} p_i\right) \sum_{i=0}^{n-1} \frac{-\ln(\prod_{i=1}^{n} p_i)}{i!}, \tag{3}$$

where P_i is the p-value obtained for evidence i, and n is the number of evidence.

A generic pathway crosstalk reference map is then built as a network, where the nodes represent pathways and the edges represent a statistically significant combined p-value for crosstalk likelihood between a pathway pair (at a significance level of $\alpha = 0.01$).

2.2. Identification of Patient-Specific Pathway Crosstalks

To determine which of the pathway crosstalks in the generic reference map may be utilized as a biomarker for MCI progression to AD dementia progression, we identify patient-specific pathway crosstalks. For this purpose, we make use of SNP data. SNPs are variations in the deoxyribonucleic acid (DNA) sequence at particular locations, which can influence phenotypes such as proneness to disease or reaction to drugs. Initiatives such as the ADNI collect patient-specific SNP information. We utilize this information to identify patient-specific pathway crosstalks via the following four steps (Figure 2):

1. Obtain a mapping of SNPs to pathways using genetic information.
2. Identify the list of SNPs that are present in a patient.
3. Use the mapping obtained in Step 1 and the patient-specific SNP list in Step 2 to obtain the pathways that are "SNP-enriched" in the patient.
4. Use the "SNP-enriched" pathways from Step 3 to obtain patient-specific pathway crosstalks.

2.2.1. Obtain a Mapping of SNPs to Pathways

Every SNP is assigned a chromosome number and a location on the genome, which can be used to map SNPs to genes, and, in turn, SNPs to pathways. Starting with a list of all genes that map to at least one pathway, we assign an SNP to a gene if it is present within 10 kilo base pairs (kbp) distance upstream or downstream of that gene. This method has been previously used by Silver et al. [36,37]. Note that since SNPs are mapped to all genes within a range of 10 kbp, the same SNP may be mapped to more than one gene. The set of SNPs assigned to a pathway is the union of all SNPs assigned to the genes of that pathway.

2.2.2. Identify Patient-Specific SNPs That Are Present

For each patient, we identify a list of SNPs that are present based on the homozygous minor (recessive) genetic model. This genetic model requires a minor allele count of 2 for an SNP to be considered present, i.e., the minor allele is inherited from both parents.

2.2.3. Identify Patient-Specific SNP-Enriched Pathways

Given the set of SNPs assigned to a pathway, $SNP_{pathway}$, the set of SNPs that are present in a patient, $SNP_{patient}$, and the set of SNPs of interest, $SNP_{interest}$, we define an enrichment score for this pathway and patient as Equation (4):

$$Enrichment(patient, pathway) = \frac{\left|SNP_{patient} \cap SNP_{pathway} \cap SNP_{interest}\right|}{\left|SNP_{interest}\right|}, \tag{4}$$

where $SNP_{interest}$ is the set of all SNPs found on the human genome or a set of relevant SNPs from the scientific literature.

A *p*-value for the enrichment score is calculated using Monte Carlo methods, as discussed previously. The "SNP-enriched" pathways for each patient are then defined as the pathways with a statistically significant *p*-value for that patient (at a significance level of $\alpha = 0.05$).

2.2.4. Identify Patient-Specific Pathway Crosstalks

Given the SNP-enriched pathways for each patient, we build patient-specific pathway crosstalk maps from the generic pathway crosstalk reference map, analogous to building organism-specific pathway maps from the KEGG pathway reference map. A pathway crosstalk, i.e., an edge in the patient-specific reference map, is present if both pathways are SNP-enriched for that patient.

2.3. Identification of Biased Pathway Crosstalk

The pathways and patient-specific pathway crosstalks that are biased towards MCI progressive patients or MCI non-progressive patients (at a significance level of $\alpha = 0.01$) are incorporated as features into the model to predict MCI progression to AD dementia progression. The bias of an active pathway crosstalk towards MCI progressive patients is quantified using the hypergeometric test (Equation (5)):

$$\phi(n,x,v,w) = \sum_{i=w}^{x} \frac{\begin{pmatrix} x \\ i \end{pmatrix} \begin{pmatrix} n-x \\ v-i \end{pmatrix}}{\begin{pmatrix} n \\ v \end{pmatrix}}, \tag{5}$$

where

- Population: n is the total number of patients.
- Success in population: x is the total number of MCI progressive patients and y is the number of MCI non-progressive patients.
- Sample: v is the total number of patients (both MCI progressive and MCI non-progressive patients) a pathway crosstalk is enriched in.
- Success in sample: w is the number of MCI progressive patients and z is the number of MCI non-progressive patients the pathway crosstalk is enriched in.

Similarly, the bias of an active pathway crosstalk towards MCI non-progressive patients can be calculated via $\phi(n,y,v,z)$.

2.4. Datasets

In this study, we utilize cellular subsystems that model biological pathways. Henceforth, we will refer to a cellular subsystem as a pathway. To create a potential pathway crosstalk reference map, we used cellular pathway data from the KEGG database [38–40]. We obtained evidence for human chemical interaction, genetic interaction, and synthetic lethal gene pairs from BioGRID [41], domain interaction from GeneMania [42], transcription factors from the FANTOM database [43,44], and protein phosphorylation [45]. We obtained SNPs associated with genes that were manually curated to be associated with AD from the Comparative Toxicogenomics Database [46], and we obtained a compilation of genes from the literature that have been identified as likely risk factors of AD from SNPedia [47]. This information was utilized as our biologically meaningful knowledge priors. Some of the genes associated with Alzheimer's that were used in this study can be found in Table 1.

Table 1. Some of the genes associated with Alzheimer's disease (AD) that were used in this study.

Gene	Evidence
APP amyloid beta (A4) precursor protein	Mutations in this gene have been implicated in autosomal dominant AD and cerebroarterial amyloidosis (NCBI Entrez Gene)
IL-1β	Four new genetic studies underscore the relevance of IL-1 to Alzheimer's pathogenesis, showing that homozygosity of a specific polymorphism in the IL-1α gene at least triples Alzheimer's risk, especially for an earlier age of onset and in combination with homozygosity for another polymorphism in the IL-1β gene [48]
SOD2	A polymorphism in SOD2 is associated with development of AD [49]
NOS3	NOS3 may be a new genetic risk factor of late onset AD [50]

The data used in the preparation of this manuscript were obtained from the ADNI [51] database. The ADNI was launched in 2003 as a public–private partnership, led by Principal Investigator Michael W. Weiner, MD. The primary goal of the ADNI has been to test whether serial MRI, PET, other biological markers, and clinical and neuropsychological assessment can be combined to measure the progression of MCI and early AD.

For our predictive study, we utilized the dataset from an earlier study by Shaffer et al. [52] based on ADNI-1. That particular study identified 97 MCI patients and predicted progression to AD dementia based on their clinical parameters, MRI results, PET scans, cerebrospinal fluid (CSF) markers (tau, *p*-tau181P, and *β*-amyloid1-42), the *APOE ε*4 genotype, and results from at least one follow-up clinical examination. Out of the 97 patients from the earlier study, only 91 patients have corresponding SNP data in the ADNI database. Hence, for the current study, we only utilized these 91 patients. However, this reduction in the number of patients did not considerably affect the ratio of MCI progressive patients to MCI non-progressive patients. The original study had 43 MCI progressive patients and 54 MCI non-progressive patients, and the reduced dataset has 41 MCI progressive patients and 50 MCI non-progressive patients. Thus, there is still sufficient representation of the two classes of patients.

3. Results/Discussion

3.1. Sample Characteristics

The mean age for all 91 MCI patients was 74.96 ± 7.32 years (mean ± standard deviation). The male-to-female ratio was 2.37, and 96.7% of subjects were white. A total of 36.26% of subjects had a family history of AD, and 54.94% had a positive finding for the *APOE ε*4 genotype. The mean follow-up duration for all of the subjects was 31.6 ± 10.6 months. Of these, 41 progressed to AD during follow-up (MCI progressive patients) and 50 did not (MCI non-progressive patients), with MCI progressive patients tending to have longer follow-up times by about 4.5 months. Statistically, MCI progressive patients did not differ from MCI non-progressive patients in mean age, sex ratio, education, race, ethnicity, family history of AD, or *APOE ε*4 prevalence. See Table 2 for details.

Table 2. Baseline Characteristics of mild cognitive impairment (MCI) Study Sample.

Subjects with MCI (*n* = 91)	MCI Progressive Patients (*n* = 41)	MCI Non-Progressive Patients (*n* = 50)	*p*-Value [1]
Age (years)	75.17 ± 7.30	74.78 ± 7.44	0.8011
Male-to-female-ratio [2,3]	2.42 (29/12)	2.33 (35/15)	0.9394
Family history of AD [2,3]	36.39 (15/41)	36.00 (18/50)	0.9539
*APOE ε*4 carriers, % [2,3]	60.98 (25/41)	50.00 (25/50)	0.4036
Average follow-up time (months)	34.10 ± 9.70	29.64 ± 10.94	0.0426

[1] Data in parentheses are number of participants. [2] *p*-values obtained using χ^2-tests. [3] Use of χ^2- or *t*-test to compare difference between MCI progressive patients and MCI non-progressive patients. Unless otherwise indicated, the data is written as mean ± standard deviation and *p*-values were calculated using a *t*-test.

3.2. SNP-Enriched Pathways and Associated Crosstalks

Our analysis identified SNP-enriched pathways that represent six of the seven KEGG pathway categories, including Cellular Processes, Metabolism, Environmental Information Processing, Genetic Information Processing, Human Diseases, and Organismal Systems. This broad array of pathway categories represents the complex nature of AD pathogenesis, which has been attributed to many different biological mechanisms, ranging from amyloid toxicity to metabolic dysfunction to immune dysregulation. Figure 3 depicts the distribution of SNP-enriched pathways amongst the six KEGG categories. The majority of enriched pathways are classified under Human Diseases (31%). This supports the well-established relationships between AD and multiple other cardiovascular, autoimmune, and neurodegenerative diseases. For instance, diabetes, obesity, and heart diseases are well-established risk factors of AD, so much so that AD has been referred to as type 3 diabetes. As such, finding SNP-enriched pathways for cardiovascular, endocrine, and metabolic diseases in individuals with MCI is anticipated [53].

Similarly, the enrichment of metabolic pathways, organismal systems including nervous and immune system pathways, and common signaling pathways of the environmental information processing category is also expected and well-supported in the literature [54–68]. Interestingly, several genetic information processing pathways, including cell cycle regulation and DNA replication and repair, were found to be enriched. Evidence for the roles of these pathways in AD has only recently begun to surface [69–71]. Our findings of the SNP-enrichment of these pathways among MCI individuals may provide support for further investigations into such pathways.

SNP-enriched pathway crosstalks were discovered between six KEGG categories, with the greatest number of crosstalks occurring between Human Diseases and Organismal systems. It is difficult to stipulate the significance of these findings. However, given that the etiology of many diseases, including AD, is complex and likely involves the failure/dysregulation of many pathways that are involved in the normal functioning of multiple organ systems, such significant crosstalk between these two categories among MCI individuals is not unexpected. The ageing process itself may facilitate a greater number of crosstalks in many pathways, since aging is associated with degeneration in many tissues and raises the risk for other chronic diseases besides dementia.

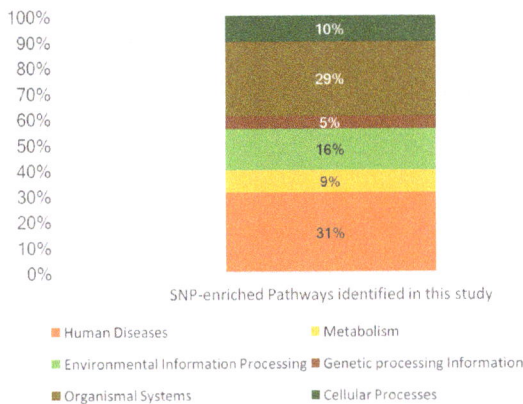

Figure 3. The distribution of the types of SNP-enriched pathways identified in this study and a comparison to the pathway distribution of the Kyoto Encyclopedia of Genes and Genomes (KEGG). NOTE: Although there are seven KEGG pathway categories, here we only show the six KEGG pathway categories that included identified SNP-enriched pathways in this study.

To investigate the genetic load in regards to AD, we further examined enriched pathway crosstalks specifically relating to the KEGG AD pathway. We identified 97 AD-related crosstalks and grouped the participating pathways by KEGG category (Figure 4).

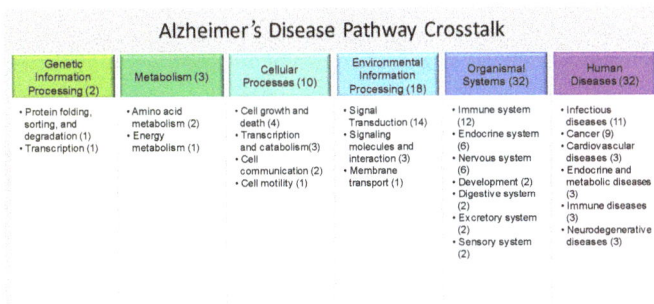

Figure 4. Pathways found to have significant crosstalk with the AD pathway and corresponding KEGG categories (shown in colored blocks). Specific KEGG pathway types are listed below each category with the number of occurrences in parentheses. NOTE: Although there are seven KEGG pathway categories, here we only show the six KEGG pathway categories that included identified SNP-enriched pathways in this study.

In line with the overall findings of crosstalk enrichment, the AD-specific pathway crosstalks primarily fell between the categories Human Diseases and Organismal Systems, supporting the importance of the pathways within these categories in AD genetic load. In contrast, pathways of Metabolism and Genetic Information Processing had very few crosstalks, suggesting that genetic load in these processes is not as important to the disease process, at least in this particular cohort. Similar findings were seen in the analysis of all pathway crosstalks. Focusing in on the AD pathway, we observe significant crosstalk in between all pathway categories supporting the complex etiology of this disease.

3.3. SNP-Enriched Features with Baseline Clinical Parameters

We predicted MCI progression to AD dementia progression using a support vector machine (SVM) with a linear kernel function with baseline clinical parameters (age, education, and Alzheimer's disease assessment scale-cognitive subscale (ADAS-Cog)), significant pathways, or significant pathway crosstalks as predictors. The results for 100 iterations of 10-fold cross-validation are shown in Table 3. The model built with the clinical parameters only produced an accuracy of $59.19 \pm 2.46\%$ with $83.64 \pm 0.29\%$ of training data points as support vectors. The model built with significant pathways alone produced an accuracy of $56.78 \pm 3.5\%$ with $68.36 \pm 3.5\%$ support vectors. Typically, we expect a random guessing model to yield an accuracy of 50%; thus, both models only perform moderately above a random model.

Table 3. Performance of support vector machine (SVM) models with baseline clinical parameters.

Metrics	Baseline Clinical: Age, Education, ADAS-Cog	Significant Pathways (Only)	Significant Pathway Crosstalks (Only)	Baseline Clinical + Significant Pathways	Baseline Clinical + Significant Pathway Crosstalks
Accuracy in %	59.19 ± 2.46	56.78 ± 3.5	60.97 ± 3.24	64.57 ± 3.56	70.9 ± 3.3
Support Vectors in %	83.64 ± 0.29	68.36 ± 2.1	50.83 ± 4.77	63.3 ± 1.15	54.29 ± 0.56
True Positives	30.78 ± 1.7	31.21 ± 4.7	40.9 ± 3.1	33.64 ± 2.4	37.93 ± 2.16
False Negatives	19.22 ± 1.7	18.79 ± 4.7	9.06 ± 3.13	16.36 ± 2.4	12.07 ± 2.16
False Positives	17.9 ± 1.6	20.51 ± 3.09	26.46 ± 4.17	15.9 ± 2.03	14.41 ± 1.6
True Negatives	23.08 ± 1.55	20.49 ± 3.09	14.54 ± 4.17	25.11 ± 2.03	26.59 ± 1.6
Sensitivity	0.62 ± 0.03	0.62 ± 0.09	0.82 ± 0.06	0.67 ± 0.05	0.76 ± 0.04
Specificity	0.56 ± 0.04	0.51 ± 0.07	0.35 ± 0.11	0.61 ± 0.05	0.65 ± 0.04
Precision	0.62 ± 0.03	0.60 ± 0.09	0.61 ± 0.03	0.68 ± 0.03	0.74 ± 0.03

ADAS-Cog, Alzheimer's disease assessment scale-cognitive subscale.

A high percentage of support vectors indicate that an SVM model is overfitted and unlikely to generalize well. Thus, if we have two models that produce the same accuracy, then we pick the model that has the lower percentage of support vectors. Sixty-eight percent (68%) or more of the training data points were used as support vectors and this indicates highly overfitted models, which is shown by the poor cross-validation accuracy.

Incorporating both the baseline clinical parameters and significant pathways as predictors produced a model with an accuracy of $64.57 \pm 3.56\%$ with $63.3 \pm 1.15\%$ support vectors. This combined model demonstrated a 5.38% increase in accuracy compared to the baseline clinical parameters model and a 7.79% increase in accuracy compared to the model using significant pathways alone. Additionally, the reduced support vector percentage of this combined model indicates a better generalizability than the baseline clinical parameters model (20.34% decrease in support vectors) and the significant pathways model (5.04% decrease in support vectors).

With our novel approach of using significant pathway crosstalks to predict AD progression, our model provides an accuracy of 60.97 ± 3.24, which is higher than using baseline clinical parameters or significant pathways alone. Furthermore, this crosstalks model has the lower support vector percentage of $50.83 \pm 4.77\%$, and thus the greatest generalizability of all of the models.

The enhancement of the significant pathway crosstalks model with the inclusion of baseline clinical parameters produced a model that has the greatest accuracy of 70.9 ± 3.3% with a moderate support vectors percentage of 54.29 ± 0.56%. These initial results support the utility of using pathway crosstalks as significant predictors in the progression from MCI progression to AD dementia and warrant replication in larger samples followed for longer periods.

3.4. Comparison of Model Performances from Shaffer et al. (2013) with Our Model Performance including SNP-Enriched Features

We compared models built using the clinical parameters and the SNP-enriched features (significant pathways or significant pathway crosstalks) to a logistic regression model with only clinical parameters by Shaffer et al. [52] (Table 4). We also noticed that the average accuracy of the logistic regression model slightly increased (from 58.7% to 59.10 ± 1.71%) when we repeatedly created random 10-folds instead of using the 10 original folds from Shaffer et al. [52]. It decreased (to 57.04 ± 2%) when we removed the six patients that did not have corresponding SNP data in the ADNI database. Our method, when incorporating either significant pathways or significant pathway crosstalks, had a higher average accuracy on 100 randomly generated 10-folds than the method by Shaffer et al. [52]. Impressively, the combination of the baseline clinical parameters, *APOE ε4*, and significant pathway crosstalks in our logistic regression model yielded an accuracy of 72.1 ± 2.66. Also, a similar accuracy was obtained using a linear kernel SVM built on the SNP-enriched features. This indicates that the pathways and pathway crosstalks indeed lead to a better rate of prediction from MCI progression to AD dementia progression.

Table 4. Performance of Shaffer et al. [52] model with clinical parameters with 97 patients in comparison to our model with 97 and 91 patients.

Model	Logistic Regression					SVM with Linear Kernel	
No. Patient	97 Patients		91 Patients				91 Patients
Metrics	original 10-fold cross-validation: Baseline Clinical + *APOE ε4* Shaffer et al. paper [52]	100 Iterations of Random 10-fold cross-validation: Baseline Clinical + *APOE ε4*	100 Iterations of Random 10-fold cross-validation: Baseline Clinical + *APOE ε4*	100 Iterations of Random 10-fold cross-validation: Baseline Clinical + *APOE ε4* + significant pathway crosstalks	100 Iterations of Random 10-fold cross-validation: Baseline Clinical + *APOE ε4* + significant pathway	100 Iterations of Random 10-fold cross-validation: Baseline Clinical + *APOE ε4* + significant pathway crosstalks
Accuracy in %	58.7	59.10 ± 1.71	57.04 ± 2	72.1 ± 2.66	63.56 ± 3.4	69.53 ± 2.9
Support Vectors in %	N/A	N/A	N/A	N/A	63.29 ± 1.16	53.65 ± 0.6
True Positives	17	39.98 ± 1.21	16.48 ± 1.16	39.22 ± 2.08	33.74 ± 2.4	37.6 ± 1.9
False Negatives	26	14.02 ± 1.21	24.43 ± 1.3	10.78 ± 2.08	16.3 ± 2.35	12.44 ± 1.9
False Positives	14	25.65 ± 1.23	14.67 ± 1.58	14.61 ± 1.3	16.91 ± 2	15.29 ± 1.5
True Negatives	40	17.35 ± 1.23	35.52 ± 1.53	26.39 ± 1.3	24.09 ± 2	25.71 ± 1.5
Sensitivity	0.40	0.74 ± 0.02	0.41 ± 0.05	0.78 ± 0.04	0.68 ± 0.05	0.75 ± 0.04
Specificity	0.74	0.40 ± 0.03	0.72 ± 0.03	0.64 ± 0.03	0.59 ± 0.05	0.63 ± 0.04
Precision	0.46	0.61 ± 0.02	0.54 ± 0.06	0.75 ± 0.03	0.68 ± 0.04	0.73 ± 0.03

3.5. Randomized SNP-Enriched Features

To demonstrate that the pathway crosstalks found in this study have true predictive power and the results are not a random occurrence, we generated 25 random samples of pathway crosstalks with no prior association to Alzheimer's and performed 100 iterations of 10-fold cross-validation for each of these 25 samples. The results are shown in Table 5.

Table 5. Performance of models with randomized pathway cross-talk features.

Metrics	Baseline Clinical + Randomized Significant Pathway Crosstalks	Baseline Clinical + Significant Pathway Crosstalks
Accuracy in %	59.27 ± 3.66	70.9 ± 3.3
Support Vectors in %	83.47 ± 1.84	54.29 ± 0.56
True Positives	30.86 ± 1.98	37.93 ± 2.16
False Negatives	19.14 ± 1.97	12.07 ± 2.16
False Positives	17.95 ± 1.59	14.41 ± 1.6
True Negatives	23.05 ± 1.56	26.59 ± 1.6
Sensitivity	0.62 ± 0.97	0.76 ± 0.04
Specificity	0.56 ± 1.45	0.65 ± 0.04
Precision	0.63 ± 0.02	0.74 ± 0.03

The model with the baseline clinical parameters and randomized significant pathway crosstalks gave an accuracy of 59.27 ± 3.66 with 83.47 ± 1.84 support vectors. This model yields 12% less accuracy and a 29.1% increase in support vectors, in comparison to the original model that uses baseline parameters and significant pathway crosstalks (instead of randomized significant pathway crosstalks). As expected, our randomly generated pathway crosstalks shows worse performance than significant pathway crosstalks. The model accuracy is still moderately above a random guessing model, likely due to the presence of the clinical parameters. A similar trend was seen when investigating models with baseline clinical parameters and all AD biomarkers to determine the effects of randomized pathways.

In this work, we focus on the development of a novel computational methodology for the discovery of pathway crosstalks to be used as biomarkers for the prognosis of AD. To demonstrate the efficacy of our methodology, we compared it with methods and results from prior studies in this area, which used ADNI-1 data. Although there is more recent data available, ADNI-1 data was used so that we could benchmark our methodology against these prior studies. In future work, we will continue our characterization efforts by incorporating the newer ADNI datasets as well as increasing the sensitivity of the proposed methodology through the use of the additive genetic model for the identification of patient-specific SNPs. There are also some limitations to our study. The ADNI is not a population-based study; it is essentially a biomarker cohort at research sites and our sample size was relatively small: we relied on a sample that was previously studied, since our initial goal was to examine the additive value of crosstalk biomarkers. We also did not incorporate other biomarkers such as tau, p-tau181P, β-amyloid1-42, *APOE* ε4, and microRNAs at this time, since our main focus was on methodological development for discovering and characterizing pathway crosstalks. However, the ADNI results have formed the basis for many current clinical prevention drug trials, and hence the ADNI is a highly relevant dataset. Moreover, its careful selection criteria and the way it makes available rich biomarker and genetic data and longitudinal cognitive data are enormous strengths. Indeed, the study of pathway crosstalks may yield novel insights into how AD pathological (e.g., beta-amyloid, tau) and neuronal loss (e.g., apoptosis, atrophy) mechanisms interact, and our methods lay the foundation for such future work.

The generic pathway crosstalk reference map was built using several different datasets, and hence the question arises as to whether all datasets should be treated equally. For simplicity, in this study, we treated all datasets equivalently. However, a modification to our information fusion method would allow us to introduce parameters to weigh evidence differently based on expert knowledge or trustworthiness. In the future, we would like to perform additional experiments to see the effects of these parameters on disease AD prognosis. This is non-trivial, as we would first need to define a weighting scheme and then develop additional methods to gauge the weights for different evidence.

4. Conclusions

AD is a major public health challenge, and there remain substantial gaps in our knowledge of its biology and treatment targets. Fully characterizing AD at a systems biology level is a priority for these reasons. In this work, we demonstrate a new methodology to build a pathway crosstalk reference map using the combined power of several gene and protein knowledge antecedents, and use this to make AD-specific discovery pathway crosstalks by enrichment with patient-specific SNP information. Our pilot data documents the promise of utilizing those SNP-enriched pathway crosstalks to identify potential AD-linked mechanisms at a systems level. More specifically, we demonstrate a three-step methodology to build a generic pathway crosstalk reference map by combining several protein/gene evidence. We then used the identified pathway crosstalks from this map as potential AD biomarkers by enriching them with patient-specific SNP information. In an initial sample of at risk subjects, we found that utilizing SNP-enriched pathway crosstalks as additional features significantly improved the prediction accuracy of MCI progression to AD dementia progression.

In addition, we verified some previously identified pathways and identified some new pathway crosstalks that warrant further study. Furthermore, we built the prediction model including the identified pathways and crosstalks, and compared our model's outputs with a previous study. These prediction model comparison analyses show that the identified pathways and crosstalks can be used as significant biomarkers of MCI progression to AD dementia progression prediction with other clinical information. Additional analysis would be required to understand the biological mechanisms that explain the association of these pathways to AD.

In summary, this is the first report to our knowledge that characterizes biological crosstalk pathways in subjects at risk of AD using gene and protein knowledge antecedents and studies their potential utility as prognostic biomarkers. Further application of this methodology to the full ADNI-1 and ADNI-2 cohort as well as to other population studies is warranted, and may yield further insights into disease mechanisms as well as novel targets for biomarker development and drug discovery.

Acknowledgments: P.M.D. acknowledges the Cure Alzheimer's Fund and Karen L. Wrenn Trust for support. Data collection and sharing for this project was funded by the Alzheimer's Disease Neuroimaging Initiative (ADNI) (National Institutes of Health Grant U01 AG024904) and DOD ADNI (Department of Defense award number W81XWH-12-2-0012). A.D.N.I. is funded by the National Institute on Aging, the National Institute of Biomedical Imaging and Bioengineering, and through generous contributions from the following: AbbVie, Alzheimer's Association; Alzheimer's Drug Discovery Foundation; Araclon Biotech; BioClinica, Inc.; Biogen; Bristol-Myers Squibb Company; CereSpir, Inc.; Eisai Inc.; Elan Pharmaceuticals, Inc.; Eli Lilly and Company; EuroImmun; F. Hoffmann-La Roche Ltd and its affiliated company Genentech, Inc.; Fujirebio; GE Healthcare; IXICO Ltd.; Janssen Alzheimer Immunotherapy Research & Development, LLC.; Johnson & Johnson Pharmaceutical Research & Development LLC.; Lumosity; Lundbeck; Merck & Co., Inc.; Meso Scale Diagnostics, LLC.; NeuroRx Research; Neurotrack Technologies; Novartis Pharmaceuticals Corporation; Pfizer Inc.; Piramal Imaging; Servier; Takeda Pharmaceutical Company; and Transition Therapeutics. The Canadian Institutes of Health Research is providing funds to support ADNI clinical sites in Canada. Private sector contributions are facilitated by the Foundation for the National Institutes of Health (www.fnih.org). The grantee organization is the Northern California Institute for Research and Education, and the study is coordinated by the Alzheimer's Disease Cooperative Study at the University of California, San Diego. ADNI data are disseminated by the Laboratory for Neuro Imaging at the University of Southern California. We would like to thank the anonymous reviewers for their insightful comments.

Author Contributions: K.P. and N.F.S. conceived and designed the computational and genetic crosstalk experiments with advice from R.E.T., A.J.S., J.R.P., and P.M.D.; P.M.D. oversaw biomarker and clinical data collection at Duke and A.J.S. headed the genetic core of A.D.N.I. K.P. performed the experiments and analyzed the data with K.N, S.H., G.B., and P.T.D; K.P. wrote the paper and D.S., K.S., and P.S.Y. did the manuscript drafting. All authors assisted with interpretation and editing.

Conflicts of Interest: All authors have received research grants and/or advisory fees from several government agencies, advocacy groups, and/or pharmaceutical/imaging companies. P.M.D. owns stock in several companies whose products are not discussed here. A.J.S. heads the genetics core of A.D.N.I. Additional support was provided to A.J.S. by N.I.H. grants P30 AG10133 and R01 AG19771. R.E.T. was supported by the Cure Alzheimer's Fund, N.I.H. 1RF1AG048080-01 and 5R37MH060009. P.M.D. and D.S. were supported by the Cure Alzheimer's Fund and the Karen L. Wrenn Trust.

References

1. Brookmeyer, R.; Evans, D.A.; Hebert, L.; Langa, K.M.; Heeringa, S.G.; Plassman, B.L.; Kukull, W.A. National estimates of the prevalence of Alzheimer's disease in the United States. *Alzheimers Dement.* **2011**, *7*, 61–73. [CrossRef] [PubMed]
2. Alzheimer's Association. 2017 Alzheimer's Disease Facts and Figures. *Alzheimers Dement.* **2017**, *13*, 325–373.
3. Heron, M. Deaths: Leading causes for 2010. *Natl. Vital Stat. Rep.* **2013**, *62*, 1–96. [PubMed]
4. Saykin, A.J.; Shen, L.; Yao, X.; Kim, S.; Nho, K.; Risacher, S.L.; Ramanan, V.K.; Foroud, T.M.; Faber, K.M.; Sarwar, N.; et al. Genetic studies of quantitative MCI and AD phenotypes in ADNI: Progress, opportunities, and plans. *Alzheimers Dement.* **2015**, *11*, 792–814. [CrossRef] [PubMed]
5. Sheinerman, K.S.; Tsivinsky, V.G.; Abdullah, L.; Crawford, F.; Umansky, S.R. Plasma microRNA biomarkers for detection of mild cognitive impairment: Biomarker validation study. *Aging Albany N.Y.* **2013**, *5*, 925–938. [CrossRef] [PubMed]
6. Galimberti, D.; Villa, C.; Fenoglio, C.; Serpente, M.; Ghezzi, L.; Cioffi, S.M.; Arighi, A.; Fumagalli, G.; Scarpini, E. Circulating miRNAs as potential biomarkers in Alzheimer's disease. *Alzheimers Dis.* **2014**, *42*, 1261–1267. [CrossRef]
7. Femminella, G.D.; Ferrara, N.; Rengo, G. The emerging role of microRNAs in Alzheimer's disease. *Front. Physiol.* **2015**, *6*. [CrossRef] [PubMed]
8. Chen, Z.; Padmanabhan, K.; Rocha, A.M.; Shpanskaya, Y.; Mihelcic, J.R.; Scott, K.; Samatova, N.F. SPICE: Discovery of phenotype-determining component interplays. *BMC Syst. Biol.* **2012**, *6*, 40. [CrossRef] [PubMed]
9. Goh, C.S.; Gianoulis, T.A.; Liu, Y.; Li, J.; Paccanaro, A.; Lussier, Y.A.; Gerstein, M. Integration of curated databases to identify genotype-phenotype associations. *BMC Genom.* **2006**, *7*, 257. [CrossRef] [PubMed]
10. Gonzalez, O.; Zimmer, R. Assigning functional linkages to proteins using phylogenetic profiles and continuous phenotypes. *Bioinformatics* **2008**, *24*, 1257–1263. [CrossRef] [PubMed]
11. Hendrix, W.; Rocha, A.M.; Elmore, M.T.; Trien, J.; Samatova, N.F. Discovery of Enriched Biological Motifs Using Knowledge Priors with Application to Biohydrogen Production. In Proceedings of the BIOCOMP, Las Vegas, NV, USA, 12–15 July 2010; pp. 17–23.
12. Hendrix, W.; Rocha, A.M.; Padmanabhan, K.; Choudhary, A.; Scott, K.; Mihelcic, J.R.; Samatova, N.F. DENSE: Efficient and prior knowledge-driven discovery of phenotype associated protein functional modules. *BMC Syst. Biol.* **2011**, *5*, 172. [CrossRef] [PubMed]
13. Jim, K.; Parmar, K.; Singh, M.; Tavazoie, S. A cross-genomic approach for systematic mapping of phenotypic traits to genes. *Genom. Res.* **2004**, *14*, 109–115. [CrossRef] [PubMed]
14. Schmidt, M.C.; Rocha, A.M.; Padmanabhan, K.; Chen, Z.; Scott, K.; Mihelcic, J.R.; Samatova, N.F. Efficient α,β-motif finder for identification of phenotype-related functional modules. *BMC Bioinform.* **2011**, *12*, 440. [CrossRef] [PubMed]
15. Schmidt, M.C.; Rocha, A.M.; Padmanabhan, K.; Shpanskaya, Y.; Banfield, J.; Scott, K.; Mihelcic, J.R.; Samatova, N.F. NIBBS-Search for fast and accurate prediction of phenotype-biased metabolic systems. *PLoS Comput. Biol.* **2012**, *8*, e1002490. [CrossRef] [PubMed]
16. Slonim, N.; Elemento, O.; Tavazoie, S. Ab initio genotype-phenotype association reveals intrinsic modularity in genetic networks. *Mol. Syst. Biol.* **2006**, *2*. [CrossRef] [PubMed]
17. Korbel, J.O.; Doerks, T.; Jensen, L.J.; Perez-Iratxeta, C.; Kaczanowski, S.; Hooper, S.D.; Andrade, M.A.; Bork, P. Systematic association of genes to phenotypes by genome and literature mining. *PLoS Biol.* **2005**, *3*, e134. [CrossRef] [PubMed]
18. Levesque, M.; Shasha, D.; Kim, W.; Surette, M.G.; Benfey, P.N. Trait-to-Gene: A computational method for predicting the function of uncharacterized genes. *Curr. Biol.* **2003**, *13*, 129–133. [CrossRef]
19. Tamura, M.; D'haeseleer, P. Microbial genotype-phenotype mapping by class association rule mining. *Bioinformatics* **2008**, *24*, 1523–1529. [CrossRef] [PubMed]
20. Zalesky, A.; Fornito, A.; Bullmore, E.T. Network-based statistic: Identifying differences in brain networks. *Neuroimage* **2010**, *53*, 1197–1207. [CrossRef] [PubMed]
21. Ballatore, C.; Lee, V.M.; Trojanowski, J.Q. Tau-mediated neurodegeneration in Alzheimer's disease and related disorders. *Nat. Rev. Neurosci.* **2007**, *8*, 663–672. [CrossRef] [PubMed]

22. Lanni, C.; Uberti, D.; Racchi, M.; Govoni, S.; Memo, M. Unfolded p53: A potential biomarker for Alzheimer's disease. *J. Alzheimers Dis.* **2007**, *12*, 93–99. [CrossRef] [PubMed]

23. Selkoe, D.J. Alzheimer's disease: Genes, proteins, and therapy. *Physiol. Rev.* **2001**, *81*, 741–766. [CrossRef] [PubMed]

24. Li, Y.; Agarwal, P.; Rajagopalan, D. A global pathway crosstalk network. *Bioinformatics* **2008**, *24*, 1442–1447. [CrossRef] [PubMed]

25. Godoy Zeballos, J.A. Signaling pathway cross talk in Alzheimer's disease. *Cell Commun. Signal.* **2014**, *12*, 23. [CrossRef] [PubMed]

26. Ramanan, V.K.; Saykin, A.J. Pathways to neurodegeneration: Mechanistic insights from GWAS in Alzheimer's disease, Parkinson's disease, and related disorders. *Am. J. Neurodegener. Dis.* **2013**, *2*, 145–175. [PubMed]

27. Myers, C.L.; Robson, D.; Wible, A.; Hibbs, M.A.; Chiriac, C.; Theesfeld, C.L.; Dolinski, K.; Troyanskaya, O.G. Discovery of biological networks from diverse functional genomic data. *Genome Biol.* **2005**, *6*, R114. [CrossRef] [PubMed]

28. Xu, Y.; Hu, W.; Chang, Z.; Duanmu, H.; Zhang, S.; Li, Z.; Yu, L.; Li, X. Prediction of human protein-protein interaction by a mixed Bayesian model and its application to exploring underlying cancer-related pathway crosstalk. *J. R. Soc. Interface* **2011**, *8*, 555–567. [CrossRef] [PubMed]

29. Liu, Z.P.; Wang, Y.; Zhang, X.S.; Chen, L. Identifying dysfunctional crosstalk of pathways in various regions of Alzheimer's disease brains. *BMC Syst. Biol.* **2010**, *4*, S11. [CrossRef] [PubMed]

30. Li, Y. Pathway crosstalk network. In *Systems Biology for Signaling Networks*; Choi, S., Ed.; Springer: New York, NY, USA, 2010; pp. 491–504.

31. Hartman, J.L.; Garvik, B.; Hartwell, L. Principles for the buffering of genetic variation. *Science* **2001**, *291*, 1001–1004. [CrossRef] [PubMed]

32. Suthers, P.F.; Zomorrodi, A.; Maranas, C.D. Genome-scale gene/reaction essentiality and synthetic lethality analysis. *Mol. Syst. Biol.* **2009**, *5*, 301. [CrossRef] [PubMed]

33. North, B.V.; Curtis, D.; Sham, P.C. A note on the calculation of empirical *p*-values from Monte Carlo procedures. *Am. J. Hum. Genet.* **2002**, *71*, 439–441. [CrossRef] [PubMed]

34. Bailey, T.L.; Gribskov, M. Combining evidence using p-values: Application to sequence homology searches. *Bioinformatics* **1998**, *14*, 48–54. [CrossRef] [PubMed]

35. Feller, W. *An Introduction to Probability Theory and Its Applications*; Wiley: New York, NY, USA, 1957.

36. Silver, M.; Janousova, E.; Hua, X.; Thompson, P.M.; Montana, G. Identification of gene pathways implicated in Alzheimer's disease using longitudinal imaging phenotypes with sparse regression. *Neuroimage* **2012**, *63*, 1681–1694. [CrossRef] [PubMed]

37. Silver, M.; Montana, G. Fast identification of biological pathways associated with a quantitative trait using group lasso with overlaps. *Stat. Appl. Genet. Mol. Biol.* **2012**, *11*. [CrossRef] [PubMed]

38. Kanehisa, M.; Goto, S. KEGG: Kyoto encyclopedia of genes and genomes. *Nucleic Acids Res.* **2000**, *28*, 27–30. [CrossRef] [PubMed]

39. Kanehisa, M.; Goto, S.; Furumichi, M.; Tanabe, M.; Hirakawa, M. KEGG for representation and analysis of molecular networks involving diseases and drugs. *Nucleic Acids Res.* **2010**, *38*, 355–360. [CrossRef] [PubMed]

40. Kanehisa, M.; Goto, S.; Hattori, M.; Aoki-Kinoshita, K.F.; Itoh, M.; Kawashima, S.; Katayama, T.; Araki, M.; Hirakawa, M. From genomics to chemical genomics: New developments in KEGG. *Nucleic Acids Res.* **2006**, *34*, 354–357. [CrossRef] [PubMed]

41. Stark, C.; Breitkreutz, B.J.; Reguly, T.; Boucher, L.; Breitkreutz, A.; Tyers, M. BioGRID: A general repository for interaction datasets. *Nucleic Acids Res.* **2005**, *34*, 535–539. [CrossRef] [PubMed]

42. Warde-Farley, D.; Donaldson, S.L.; Comes, O.; Zuberi, K.; Badrawi, R.; Chao, P.; Franz, M.; Grouios, C.; Kazi, F.; Lopes, C.T.; et al. The GeneMANIA prediction server: Biological network integration for gene prioritization and predicting gene function. *Nucleic Acids Res.* **2010**, *38*, W214–W220. [CrossRef] [PubMed]

43. Kawaji, H.; Severin, J.; Lizio, M.; Forrest, A.R.; van Nimwegen, E.; Rehli, M.; Schroder, K.; Irvine, K.; Suzuki, H.; Carninci, P.; et al. Update of the FANTOM web resource: From mammalian transcriptional landscape to its dynamic regulation. *Nucleic Acids Res.* **2010**, *39*, 856–860. [CrossRef] [PubMed]

44. Kawaji, H.; Severin, J.; Lizio, M.; Waterhouse, A.; Katayama, S.; Irvine, K.M.; Hume, D.A.; Forrest, A.R.R.; Suzuki, H.; Carninci, P.; Hayashizaki, Y.; Daub, C.O. The FANTOM web resource: From mammalian transcriptional landscape to its dynamic regulation. *Genome Biol.* **2009**, *10*, R40. [CrossRef] [PubMed]

45. Li, T.; Du, P.; Xu, N. Identifying human kinase-specific protein phosphorylation sites by integrating heterogeneous information from various sources. *PLoS ONE* **2010**, *5*, e15411. [CrossRef] [PubMed]

46. Davis, A.P.; Murphy, C.G.; Johnson, R.; Lay, J.M.; Lennon-Hopkins, K.; Saraceni-Richards, C.; Sciaky, D.; King, B.L.; Rosenstein, M.C.; Wiegers, T.C.; et al. The comparative toxicogenomics database: Update 2013. *Nucleic Acids Res.* **2013**, *41*, 1104–1114. [CrossRef] [PubMed]

47. Cariaso, M.; Lennon, G. SNPedia: A wiki supporting personal genome annotation, interpretation and analysis. *Nucleic Acids Res.* **2012**, *40*, 1308–1312. [CrossRef] [PubMed]

48. Mrak, R.E.; Griffin, S. Interleukin-1, neuroinflammation, and Alzheimer's disease. *Neurobiol. Aging* **2001**, *22*, 903–908. [CrossRef]

49. Wiener, H.W.; Perry, R.T.; Chen, Z.; Harrell, L.E.; Go, R.C. A polymorphism in *SOD2* is associated with development of Alzheimer's disease. *Genes Brain Behav.* **2007**, *6*, 770–775. [CrossRef] [PubMed]

50. Dahiyat, M.; Cumming, A.; Harrington, C.; Wischik, C.; Xuereb, J.; Corrigan, F.; Breen, G.; Shaw, D.; St Clair, D. Association between Alzheimer's disease and the *NOS3* gene. *Ann. Neurol.* **1999**, *46*, 664–667. [CrossRef]

51. Weiner, M.W.; Veitch, D.P.; Aisen, P.S.; Beckett, L.A.; Cairns, N.J.; Green, R.C.; Harvey, D.; Jack, C.R.; Jagust, W.; Liu, E.; et al. The Alzheimer's disease neuroimaging initiative: A review of papers published since its inception. *Alzheimers Dement.* **2012**, *8*, 1–68. [CrossRef] [PubMed]

52. Shaffer, J.L.; Petrella, J.R.; Sheldon, F.C.; Choudhury, K.R.; Calhoun, V.D.; Coleman, R.E.; Doraiswamy, P.M. Predicting cognitive decline in subjects at risk for Alzheimer disease by using combined cerebrospinal fluid, MR imaging, and PET biomarkers. *Radiology* **2013**, *266*, 583–591. [CrossRef] [PubMed]

53. Vignini, A.; Giulietti, A.; Nanetti, L.; Raffaelli, F.; Giusti, L.; Mazzanti, L.; Provinciali, L. Alzheimer's disease and diabetes: New insights and unifying therapies. *Curr. Diabetes Rev.* **2013**, *9*, 218–227. [CrossRef] [PubMed]

54. Ramanan, V.K.; Kim, S.; Holohan, K.; Shen, L.; Nho, K.; Risacher, S.L.; Foroud, T.M.; Mukherjee, S.; Crane, P.K.; Aisen, P.S.; et al. Genome-wide pathway analysis of memory impairment in the Alzheimer's disease neuroimaging initiative (ADNI) cohort implicates gene candidates, canonical pathways, and networks. *Brain Imaging Behav.* **2012**, *6*, 634–648. [CrossRef] [PubMed]

55. Roe, C.M.; Behrens, M.I.; Xiong, C.; Miller, J.P.; Morris, J.C. Alzheimer disease and cancer. *Neurology* **2005**, *64*, 895–898. [CrossRef] [PubMed]

56. Sardi, F.; Fassina, L.; Venturini, L.; Inguscio, M.; Guerriero, F.; Rolfo, E.; Ricevuti, G. Alzheimer's disease, autoimmunity and inflammation. The good, the bad and the ugly. *Autoimmun. Rev.* **2011**, *11*, 149–153. [CrossRef] [PubMed]

57. Honjo, K.; van Reekum, R.; Verhoeff, N.P. Alzheimer's disease and infection: Do infectious agents contribute to progression of Alzheimer's disease? *Alzheimers Dement.* **2009**, *5*, 348–360. [CrossRef] [PubMed]

58. Jung, B.K.; Pyo, K.H.; Shin, K.Y.; Hwang, Y.S.; Lim, H.; Lee, S.J.; Moon, J.H.; Lee, S.H.; Suh, Y.H.; Chai, J.Y.; et al. Toxoplasma gondii infection in the brain inhibits neuronal degeneration and learning and memory impairments in a murine model of Alzheimer's disease. *PLoS ONE* **2012**, *7*, e33312. [CrossRef] [PubMed]

59. Al-Mansoori, K.M.; Hasan, M.Y.; Al-Hayani, A.; El-Agnaf, O.M. The role of α-synuclein in neurodegenerative diseases: From molecular pathways in disease to therapeutic approaches. *Curr. Alzheimer Res.* **2013**, *10*, 559–568. [CrossRef] [PubMed]

60. Mandrekar-Colucci, S.; Karlo, J.C.; Landreth, G.E. Mechanisms underlying the rapid peroxisome proliferator-activated receptor-γ-mediated amyloid clearance and reversal of cognitive deficits in a murine model of Alzheimer's disease. *J. Neurosci.* **2012**, *32*, 10117–10128. [CrossRef] [PubMed]

61. Berntorp, K.; Frid, A.; Alm, R.; Fredrikson, G.N.; Sjöberg, K.; Ohlsson, B. Antibodies against gonadotropin-releasing hormone (GnRH) in patients with diabetes mellitus is associated with lower body weight and autonomic neuropathy. *BMC Res. Notes* **2013**, *6*, 329. [CrossRef] [PubMed]

62. Giunta, B.; Fernandez, F.; Nikolic, W.V.; Obregon, D.; Rrapo, E.; Town, T.; Tan, J. Inflammaging as a prodrome to Alzheimer's disease. *J. Neuroinflamm.* **2008**, *5*, 51. [CrossRef] [PubMed]

63. Liu, Y.H.; Zeng, F.; Wang, Y.R.; Zhou, H.D.; Giunta, B.; Tan, J.; Wang, Y.J. Immunity and Alzheimer's disease: Immunological perspectives on the development of novel therapies. *Drug Discov. Today* **2013**, *18*, 1212–1220. [CrossRef] [PubMed]

64. Freo, U.; Pizzolato, G.; Dam, M.; Ori, C.; Battistin, L. A short review of cognitive and functional neuroimaging studies of cholinergic drugs: Implications for therapeutic potentials. *J. Neural Transm.* **2002**, *109*, 857–870. [CrossRef] [PubMed]

65. Tamburri, A.; Dudilot, A.; Licea, S.; Bourgeois, C.; Boehm, J. NMDA-Receptor activation but not ion flux is required for amyloid-beta induced synaptic depression. *PLoS ONE* **2013**, *8*, e65350. [CrossRef] [PubMed]

66. Chen, Q.S.; Wei, W.Z.; Shimahara, T.; Xie, C.W. Alzheimer amyloid beta-peptide inhibits the late phase of long-term potentiation through calcineurin-dependent mechanisms in the hippocampal dentate gyrus. *Neurobiol. Learn. Mem.* **2002**, *77*, 354–371. [CrossRef] [PubMed]

67. Ghebranious, N.; Mukesh, B.; Giampietro, P.F.; Glurich, I.; Mickel, S.F.; Waring, S.C.; Mc-Carty, C.A. A pilot study of gene/gene and gene/environment interactions in Alzheimer disease. *Clin. Med. Res.* **2011**, *9*, 17–25. [CrossRef] [PubMed]

68. Mosch, B.; Morawski, M.; Mittag, A.; Lenz, D.; Tarnok, A.; Arendt, T. Aneuploidy and DNA replication in the normal human brain and Alzheimer's disease. *J. Neurosci.* **2007**, *27*, 6859–6867. [CrossRef] [PubMed]

69. Yurov, Y.B.; Vorsanova, S.G.; Iourov, I.Y. The DNA replication stress hypothesis of Alzheimer's disease. *Sci. World J.* **2011**, *11*, 2602–2612. [CrossRef] [PubMed]

70. Yang, Y.; Geldmacher, D.S.; Herrup, K. DNA replication precedes neuronal cell death in Alzheimer's disease. *J. Neurosci.* **2001**, *21*, 2661–2668. [PubMed]

71. Bradley, W.G.; Polinsky, R.J.; Pendlebury, W.W.; Jones, S.K.; Nee, L.E.; Bartlett, J.D.; Hartshorn, J.N.; Tandan, R.; Sweet, L.; Magin, G.K. DNA repair deficiency for alkylation damage in cells from Alzheimer's disease patients. *Prog. Clin. Biol. Res.* **1989**, *317*, 715–732. [PubMed]

MDPI

St. Alban-Anlage 66

4052 Basel

Switzerland

Tel. +41 61 683 77 34

Fax +41 61 302 89 18

www.mdpi.com

Processes Editorial Office

E-mail: processes@mdpi.com

www.mdpi.com/journal/processes

www.ingramcontent.com/pod-product-compliance
Lightning Source LLC
Chambersburg PA
CBHW051900210326

41597CB00033B/5963